BASIC water and wastewater treatment

Butterworths BASIC Series includes the following titles:

BASIC aerodynamics
BASIC artificial intelligence
BASIC business analysis and operations research
BASIC business systems simulation
BASIC differential equations
BASIC digital signal processing
BASIC economics
BASIC electrotechnology
BASIC forecasting techniques
BASIC fluid mechanics
BASIC graph and network algorithms
BASIC heat transfer
BASIC hydraulics
BASIC hydrodynamics
BASIC hydrology
BASIC interactive graphics
BASIC investment appraisal
BASIC materials studies
BASIC matrix methods
BASIC mechanical vibrations
BASIC molecular spectroscopy
BASIC numerical mathematics
BASIC operational amplifiers
BASIC reliability engineering analysis
BASIC soil mechanics
BASIC statistics
BASIC stress analysis
BASIC surveying
BASIC technical systems simulation
BASIC theory of structures
BASIC thermodynamics and heat transfer
BASIC water and wastewater treatment

BASIC water and wastewater treatment

T H Y Tebbutt BSc, SM, PhD, CEng, FICE, FIWEM
Senior Lecturer, School of Civil Engineering, University of Birmingham, England

Butterworths
London Boston Singapore Sydney Toronto Wellington

First published 1990

British Library Cataloguing in Publication Data
Tebbutt, T. H. Y. (Thomas Hugh Yelland), 1935–
BASIC water and wastewater treatment.
1. Water supply. Waste water. Water supply & waste water. Treatment. Applications of computer systems. Programming languages
I. Title
628.1'62'02855133

ISBN 0-408-70937-5

Library of Congress Cataloging-in-Publication Data
Tebbutt, T. H. Y.
BASIC water and wastewater treatment / T. H. Y. Tebbutt.
p. cm. – (Butterworths BASIC series)
Includes bibliographical references.
ISBN 0-408-70937-5:
1. Water–Purification–Data processing. 2. Sewage–Purification–Data processing. 3. BASIC (Computer program language) I. Title. II. Series.
TD433. T43 1990
628.1'62–dc20

89-23948
CIP

Photoset by KEYTEC, Bridport, Dorset
Printed and bound by Hartnolls Ltd, Bodmin, Cornwall

Preface

This book is one of an expanding series which aims to combine the applications of simple programming in BASIC with an understanding of some aspects of a particular branch of engineering or science. It is not intended to be an instruction manual for programming in BASIC, nor is it a comprehensive coverage of water and wastewater treatment. An introductory chapter provides sufficient information about programming in BASIC to enable the reader to follow the programs in the succeeding chapters but is not a substitute for the numerous texts about BASIC. Water and wastewater treatment is an important area of engineering which is dealt with to some extent in most civil engineering courses and which also appears in chemical engineering, environmental engineering and environmental sciences courses. It is an area which combines aspects of engineering, biology, chemistry and physics, and the topics covered in the book have been selected to illustrate the multidisciplinary aspects of the subject. It must be appreciated that the discipline of water and wastewater treatment, dealing as it does with natural raw materials and the waste products of civilization, is not perhaps so mathematically dependent as other branches of engineering. The random variations in the composition of waters and wastewaters mean that strictly theoretical explanations of phenomena are not always of practical value and reliance may have to be placed, at least in part, on empirical and philosophical interpretations of what is observed to occur. The coverage of water and wastewater treatment in this book is thus concerned with those aspects of the subject which can be expressed by mathematical relationships or which require a decision-making sequence. For a full understanding of water and wastewater treatment, water pollution control and other environmental aspects of water, the reader is referred to the recommendations for further reading at the end of each chapter.

The choice of BASIC language for this series of books has been influenced by the enormous growth in the availability of personal computers in educational establishments and in industry. Although most of these machines will run a wide range of languages, BASIC is widely established as an easily learnt and flexible language. Whilst not ideal for complex mathematical calculations, it is capable of undertaking, reasonably effectively, most of the straightforward computing operations which are required in general engineering.

Experience with both undergraduate and postgraduate students suggests that, although there is a good deal of talk about computer awareness in schools and in undergraduate courses, many students are very uncertain when asked to write even simple programs for a real engineering problem. This may well be due to lack of keyboard experience or to the frustrations which often seem to occur when using large mainframe computers. Such lack of confidence can usually be remedied by exposure to simple BASIC programming in a personal computer environment which appears less intimidating and more friendly than mainframes using more complex languages. It is hoped that by working through this book the reader will find that computing can be useful in solving engineering problems and may even be enjoyable!

The programs in this book have been written in GWBASIC on an Amstrad PC1640 and should operate on all PC-compatible machines. I have tried to ensure that all the programs perform as intended but would not claim that they are necessarily the most efficient or the most elegant solutions to the particular requirements. I apologize in advance for any errors which I have missed in the listings, although it should be emphasized that the program listings are direct computer outputs from working programs and do not therefore contain any typographical errors.

The illustrations in the text were produced by Mrs Mary Andrews using a Macintosh II system and I am most grateful for her contribution to this book.

I hope that having progressed through the book the reader does feel that understanding of both BASIC programming and water and wastewater treatment has been improved.

T. H. Y. Tebbutt
Birmingham 1989

Contents

Chapter 1

Introduction to BASIC

1.1 BASIC language

The programs in this book are written in BASIC (Beginners All-purpose Symbolic Instruction Code), which was developed at Dartmouth College, USA, many years ago as an easy-to-learn language particularly suitable for teaching and general use. It was developed in the days of mainframe computers but because of its simplicity it was an obvious choice for use when personal computers came into existence. Its simple concepts and flexibility have thus introduced computer programming to millions of people around the world.

Inevitably, a general-purpose language is unlikely to be ideal for all specific applications and therefore in many specialized areas of computing other languages such as FORTRAN, Pascal, etc., are preferred. The simplicity of BASIC and its lack of structure are also seen as disadvantageous in some applications. Nevertheless BASIC is able to carry out most programming operations and it is likely to remain popular as an introductory language. The initial BASIC language has been subject to many variations since it was first invented; some of these changes simplify programming or speed up the running of programs, others are simply dialects for a particular machine. Problems can thus arise when a program written in the dialect for one machine is transferred to another machine which uses a different subset of BASIC. To avoid this difficulty it is now usual to write programs in a standard BASIC such as BASICA or GWBASIC as used in most PC machines. All the programs in this book are written in compatible BASIC for use on PCs and most other modern microcomputers with an integral BASIC chip. This book cannot take the place of a good manual on BASIC language and programming, several of which are listed in the Further Reading section at the end of the chapter.

A BASIC program consists of statements in numbered lines which are acted upon by the computer in sequential order.

Lines may:

assign values to a variable, e.g.

```
20 X=20
```

carry out a mathematical operation, e.g.

```
30 Y=Z*3.5
```

ask for instructions, e.g.

```
40 PRINT"Would you like another calculation (Y/N)?"
```

make a decision, e.g.

```
50 IF A=B THEN X=Y
```

Most dialects of BASIC allow multiple statements in a line using a colon as a separator:

```
60 IF A=B THEN X=Y:PRINT X
```

It will be apparent from these examples that BASIC statements are written in something like plain English, although the statements must be written in a carefully prescribed manner. Readers new to computing will soon find, by bitter experience, the importance of typing in program lines *exactly* as shown in the listings in later chapters. In the same context, when writing new programs it is vital that the proper syntax and format be adhered to exactly. For example, the difference in effect between a comma and a semicolon when displaying information is considerable, as will be explained later.

1.2 Variables

Variables are items in a program to which a value can be assigned and whose value may change during the running of a program. In standard BASIC, simple variables must be identified by a letter, A; two letters, AA; or a letter and a number, A1. With some computers A is a different variable from a, and some computers recognize words as variables, e.g. VELOCITY. Care must be taken when using words as BASIC variables because some computers will only recognize the first two letters of the word, so that ALKALINITY would be taken to be the same variable as ALUMINIUM.

When text is required as part of a program, string variables must be employed. These are indicated by the symbol $ after the name of the variable and can usually be up to 255 characters in length

```
10 A$="This is a string variable"
20 PRINT A$
```

would display

```
This is a string variable
```

1.3 Input

In many situations it is necessary for the program to be provided with information by the user, about the values to be assigned to variables for example. This can be achieved by means of the INPUT statement which places a '?' on the screen and waits for the user to provide the necessary information:

```
50 INPUT X,Y,Z
```

The user must then type in the appropriate values for the variables X, Y and Z, e.g. 23, 3.12, 7. The values must be separated by commas, since otherwise the program will assume that X = 233.127 and request the values for Y and Z. The number of values entered must correspond to the number in the INPUT statement. If this is not so, the program will automatically return to the INPUT line with a 'redo' instruction. It is often useful to amplify an INPUT statement with a prompt which identifies the values required:

```
60 INPUT"Enter length, width, depth"; X,Y,Z
```

An alternative method of providing information for a program is to use data input. This involves a READ statement and a DATA line:

```
50 READ X,Y,Z
90 DATA 23,3.12,7
```

The DATA line, which can be placed anywhere in the program, must contain the same number of items as in the READ statement and must start with the word DATA. Both simple and string variables can be placed in a READ statement but the DATA line must contain the relevant variable types in the same order.

1.4 Output

To obtain the results of a program run it is necessary to instruct the computer to display them because unless this is done the machine will simply keep them to itself. The information is obtained by means of a PRINT statement:

```
80 PRINT X,Y,Z
```

This would print the numerical values of the variables X, Y and Z. Because the variables are separated by commas, the values will appear across the screen with a spacing of about one-quarter of the screen width. If a semicolon is used, the values will be separated by only a single space. If a semicolon is placed at the end of a list for printing, the line feed will be suppressed. A PRINT statement on its own with no list will produce a blank line.

It is important to note the significance of notation in PRINT statements since

```
80 PRINT "X","Y","Z"
```

will print the characters X, Y and Z spaced across the screen. Thus the statement PRINT A means print the value of the variable A, whereas the statement PRINT "A" means print the character A. It is possible, and indeed very useful, to combine text and variables:

```
90 PRINT"Width of tank=";W;"metres"
```

Material inside " " is printed as text exactly as it appears and a variable separated by semicolons causes the current value of that variable to be printed. It is also possible to carry out an arithmetic operation as part of a PRINT command:

```
50 PRINT"Area of tank is";L*W;"metres"
```

Output is normally to the VDU screen but can be converted to hard copy if a printer is available. The statements for hard copy output differ somewhat between various machines but a common form is LPRINT, which is substituted for PRINT statements.

1.5 Mathematical expressions

All dialects of BASIC contain standard mathematical operators which are carried out in the following order of precedence:

^ raising to a power
* multiply and / divide
+ add and − subtract

Thus multiplication is carried out before addition, and for operators having the same level of precedence the working is carried out from left to right. Brackets around terms force the operations within the bracket to be carried out before other operations in an expression. It is thus important to ensure that brackets are used where necessary in equations to give the correct evaluation of an expression, e.g.

$$w = 2a + 3(b + 4c)/2a$$

would be written in BASIC as

```
50 W=(2*A+3*(B+4*C))/(2*A)
```

Examples of standard mathematical functions included in BASIC are:

EXP(A) gives e to the power A
LOG(A) gives $\log_e$ of A
SQR(A) gives square root of A

Very small or very large numbers are represented by an exponential format

7.23E06 means 7.23×10^6
7.23E–06 means 7.23×10^{-6}

1.6 Conditional statements

A valuable feature of a programming language is the ability to incorporate statements which direct the program to take a particular action if, and only if, a specified condition is fulfilled. Statements of this type take the form

```
40 IF A=B THEN X=Y
```

Conditional operators can be

=	equal
<>	not equal
<	less than
>	greater than
<=	less than or equal to
>=	greater than or equal to

The conditional operators can be used with a variety of statements to control the action of a program:

print a comment or result

```
50 IF A<B THEN PRINT "Value too low!"
```

cause the program to skip unwanted lines

```
70 IF A>B THEN 150
```

bring the run to an end

```
90 IF A<=B THEN END
```

Another form of conditional control in a program is the use of the ON THEN statement. This provides multiple branching depending upon the integer value of an expression or variable and is useful when the user selects a particular situation from a menu of several and it is necessary to direct the program to the appropriate section:

```
 60 INPUT"Enter 1 for E. coli, 2 for polio, 3 for virus";S
 70 ON S GOTO 80,90,100
 80 TC=0.24
 90 TC=1.20
100 TC=6.30
```

The integer value of S must be at least 1 and cannot exceed the number of line numbers referenced in the ON THEN line.

1.7 Loops

In many mathematical problems it is often necessary to repeat a calculation with different values of a variable and this is best undertaken by means of a loop which causes the program to cycle through a series of lines, or a routine, for a specified number of times:

```
50 FOR A=1 TO 10
60 B=(A+6)^2
70 PRINT B
80 NEXT A
```

This FOR NEXT loop calculates the value of the variable B for values of the variable A from 1 to 10. When A has reached the maximum value specified in line 50, the program will continue past line 80. The increment in value of A is taken as 1 unless line 50 is modified with a STEP statement as in

```
50 FOR A=1 TO 10 STEP 0.5
```

where the increment can be positive or negative depending on whether the limits increase or decrease.

Loops can also be created by a GOTO statement but unless care is taken in their use the program may cycle indefinitely:

```
80 GOTO 50
```

will always return the program to line 50 regardless of the

number of repeats which the user wanted. The GOTO command can, however, be useful when combined with a conditional statement to direct the program to a particular routine:

```
70 IF X>A+B GOTO 300
```

The entry of CTRL C from the keyboard is usually able to stop the running of a program and enable it to break out of endless loops if these have inadvertently been included in a program.

1.8 Subroutines

Some problems require a particular expression to be evaluated several times at different stages in the run; for example, it might be necessary to calculate the volume of a tank on several occasions. Rather than repeating the equation each time it is needed, it can be written once as a subroutine and control transferred to the line which starts the subroutine as required:

```
50 INPUT"Enter length, width and depth";L,W,D
60 GOSUB 300

300 V=L*W*D
310 RETURN
```

The RETURN statement returns the program to the line following that in which the GOSUB statement occurs.

1.9 Errors and checks

It is almost inevitable that when first written, programs will contain errors. These may be in the grammar or syntax, or simply typing mistakes. BASIC usually detects syntax errors and reports them to the screen, although not always in an easily understandable form. The offending line can then be corrected by editing, or rewritten if the errors are major ones. More difficult to find are errors which occur when the same variable is given two different assignments in separate lines of a program or where a variable is used without it having been assigned a value.

In some programs it may be possible for a variable to be assigned a value of zero which is perfectly correct but which can cause the program to fail at a later line because that particular variable appears as the divisor in an expression. In such a case the screen will report 'DIVIDE BY 0 ERROR' and the program will fail. This can be avoided by including an ON ERROR statement which, if placed before the line in which the error may occur, will direct the program to an appropriate alternative line.

It is very difficult to make programs completely foolproof in use and to do so often makes the programs long and confusing because of the need to cater for every possible eventuality. Nevertheless, it is always good practice to make programs as user-friendly as possible by providing program notes, as in this book, or by using remark statements to document the program. When REMARK or REM appears in a line, the material following this statement is ignored by the computer so that it can be used as an explanatory note:

```
20 REM lines 30–120 carry out a least squares analysis
```

During the running of a program the screen display will often become full of information and a more easily interpreted display can be obtained by removing unwanted material using the CLS statement:

```
70 CLS
80 PRINT"Results of Calculations"
```

(Some versions of BASIC use a different statement to clear the screen.)

In parts of a program which interact with the user, it is helpful to warn the user if the response is invalid, since otherwise the program will simply refuse to proceed further:

```
130 INPUT"Another set of results (Y/N)";Q$
140 IF Q$="Y" OR Q$="y" THEN 10
150 IF Q$<>"N" or Q$<>"n" THEN 170
160 END
170 PRINT"Enter Y or N only . . . you entered";Q$
180 GOTO 130
```

If the user keys in capital or lower case Y the program returns for another calculation, if N or n is entered the program ends, and if any other letter is entered the user is warned of the error and returned to the question.

When writing a program it is essential that it is fully checked using actual data and that the results of the calculations are compared with those obtained from the same data by another form of calculation. It is also necessary to ensure that if invalid data are entered, e.g. negative values for data which can only take positive values in real life, the program either fails or gives results which are obviously wrong. A program which accepts invalid data and gives answers which are wrong, but not obviously so, is very dangerous. Validation of programs is therefore highly important.

FURTHER READING

The manuals provided with all computers usually give some information about BASIC programming and there are many publications which cover BASIC in more detail; the following books are a small example of what is available.

Dvorak, D. and Mussett, A., *BASIC in Action*, Butterworths, 1984

Monro, D. M., *Interactive Computing with BASIC*, Arnold, 1974.

Quillinan, A. J., *Better BASIC*, Newnes, 1985.

Chapter 2

Elements of water and wastewater treatment

Safe drinking water and adequate sanitation are primary requirements for a good standard of living and it is an unfortunate fact that millions of people throughout the world lack these essentials. It has been estimated by the World Health Organization that 80% of illnesses in developing countries are water-related and it is known that about five million babies die each year from the combined effects of unsafe water supplies and inadequate or non-existent sanitation.

Water is, of course, essential for all forms of life, and a reliable and safe supply is a necessary prerequisite for a healthy, permanent community. In sparsely populated areas it may be possible for the local streams and lakes to remain largely unpolluted but as communities grow in size so do the pressures on the environment. The importance of safe water supply and adequate sanitation was recognized by early civilizations in India, the Middle East and Crete. The Romans were expert water engineers; aqueducts brought fresh water into the centre of Rome to satisfy the hygienic and aesthetic needs of the population, and sewers took away the wastewater. The fall of the Roman Empire brought an end to these infrastructure provisions and in the Middle Ages most towns in Europe were decidedly unhealthy places in which to live and work. With the coming of the Industrial Revolution, the rapid urbanization of countries like England placed enormous demands on the almost non-existent provisions for water supply and sanitation. As a result the general living standards and life expectancy of all but the wealthy urban dwellers were very poor. Water-related diseases such as cholera and typhoid were rampant in Europe in the middle of the last century. It was in 1854 that Dr John Snow demonstrated the connection between sewage contamination of water sources and the spread of cholera by his investigation of the Broad Street Pump Outbreak in London. It was then appreciated that protection of water sources, effective water treatment and distribution systems, well-maintained sewers, comprehensive wastewater treatment facilities, and rational pollution control measures were vital factors in improving the health of a community. With this appreciation, major investments were made in public health engineering works throughout the industrialized countries towards the end of the last century and in the first half of this century. Thus, in developed countries water-related diseases are now almost unknown. Unfortunately, in many other parts of the world conditions exist today which are analogous to those in Europe last century, with the added problem of high temperatures which encourage the spread of many infections. There is thus a great need to implement appropriate water and wastewater treatment systems all over the world and this requires engineers and scientists to have a sound understanding of the main processes involved.

2.1 The water cycle

Water present in the earth's hydrosphere, although relatively small in amount, takes part in a very active natural recirculation, the hydrological cycle depicted in Figure 2.1. Solar energy provides the driving force for the various reactions which occur in the cycle, and the nature and magnitude of the forces involved are such that human control of the hydrological cycle is not possible. Much of the water in the hydrosphere is located in the oceans, in ice caps and in deep rock formations so that only about 0.6% of it is usable freshwater. The available stock of freshwater is about four million cubic kilometres, which if spread evenly over the land surface would provide a depth of some 30 metres. On average, therefore, there is a great deal of water available to satisfy the various demands. In practice, however, the distribution of water resources is very uneven, so

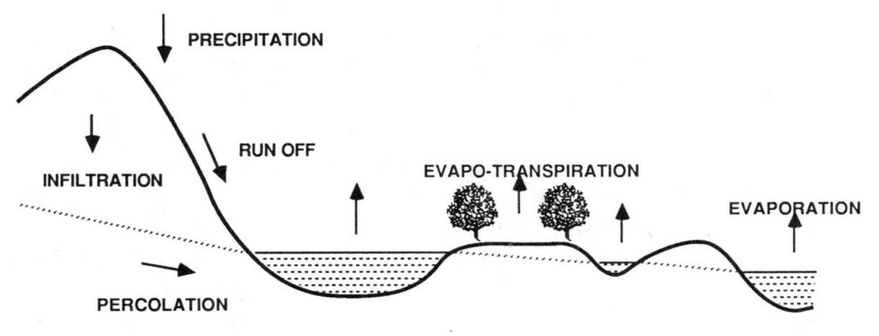

Figure 2.1 The hydrological cycle

that there are many areas of the world with large populations but with very limited availability of water. On the other hand, regions such as tropical rain forests have a great excess of water over the local demands. Rainfall, which is the basic input to water resources, is generally greatest in tropical mountainous areas, whereas large communities usually develop in relatively level sites, often close to the sea or large rivers. This means that there are many parts of the world where the natural water resources are insufficient to satisfy the domestic, industrial and agricultural demands without the intervention of water engineers and scientists. In a heavily populated environment there are many demands on water resources and there can be conflicts between various uses. In such circumstances there is a need for effective means of controlling and improving water quality, both in relation to water treatment and in wastewater treatment and pollution control, and Figure 2.2 illustrates some of the features of the engineered water cycle.

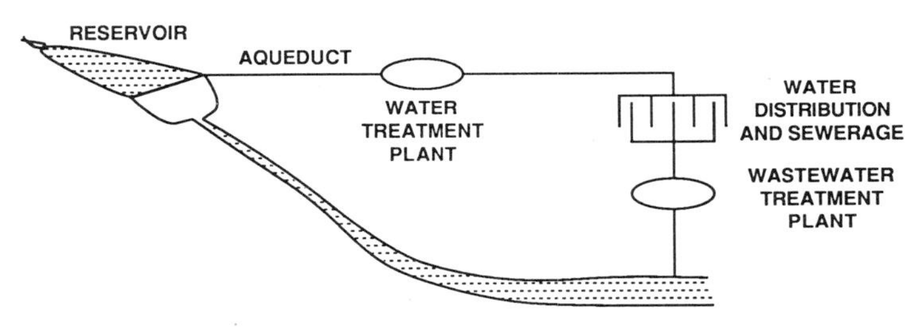

Figure 2.2 An engineered water cycle

2.2 Water quality

Because water is an almost universal solvent, all natural waters contain small amounts of other materials, dissolved from the air in the case of precipitation, and from the earth's surface and from the soil and rocks for surfacewaters and groundwaters respectively. This means that in reality there is no such thing as 'pure water', and indeed pure water with no dissolved solids or gases is actually a rather unpalatable liquid. As well as constituents, which are mainly inorganic in nature, arising from precipitation and solution of soil and rocks, there will also be a gradual accumulation of organic material from vegetation and animal wastes. The presence of organic matter in solution and in suspension will encourage the growth of microorganisms, and a natural watercourse is a complex ecological system. Human activities in a catchment result in the production of domestic and industrial wastewaters which, if not effectively treated, can destroy the ecological balance in the waters to which they are discharged. This may mean that lakes and rivers become unsuitable for many purposes such as water abstraction, fishing and recreation. Groundwaters may become contaminated by the fertilizers used in agriculture (the current nitrate problems are an example of this) or by seepage from solid waste disposal sites as well as by accidental spillages of toxic chemicals. In an area where water is in short supply, it is obviously important that as many uses as possible are satisfied by an individual resource. In developed countries there are many concerns about increasing pressures on the environment, of which water forms an important part. These pressures are just as strong in many developing countries and, although environmental protection does not perhaps receive wide attention at the moment, it is growing in importance. In assessing the quality of water and the nature of wastewaters it is necessary to utilize standard parameters to express the complex composition of water in its various natural and polluted forms.

2.3 Water quality parameters

Since water is such a good solvent there is an almost endless list of materials which could be present in a particular sample and it is not normally feasible to analyse for all possible constituents.

When assessing water quality it is therefore often convenient to use what are termed 'blanket parameters' which measure the presence of a group of contaminants or indicate a particular property. The relevance of various water quality parameters depends upon the nature of the water or wastewater and its actual or potential use but there are three basic types of characteristics which are of importance.

Physical characteristics

These are properties which are often apparent to the casual observer, and include parameters such as colour, taste, odour, temperature, suspended solids, etc.

Chemical characteristics

These include parameters such as alkalinity, hardness, organic content, nutrients, dissolved oxygen, inhibitory and toxic compounds, etc.

Biological characteristics

Natural waters normally form a balanced ecosystem containing microorganisms such as bacteria, protozoa and algae. Microorganisms provide food for fish and other higher life forms. Wastewaters often contain large numbers of microorganisms, particularly bacteria, and are potentially hazardous because of the connection with water-related diseases. Although on occasion individual species may be identified, it is common for blanket determinations of biological populations to be made, at least in the first instance.

Table 2.1 gives some examples of the characteristics of typical water and wastewater samples in terms of the commonly used parameters and also includes information relating to quality standards.

Table 2.1 Typical water quality characteristics

Parameter concn	*Upland catchment*	*Lowland river*	*Limestone aquifer*	*EEC Drinking water MAC**	*Raw sewage*
pH (units)	4	7	7.5	9.5	7
Total Solids	50	150	300	1500	1000
Chloride	5	50	25	200	150
Alkalinity	20	125	250	–	200
Colour (°H)	50	20	1	20	–
Amm. N	0.05	0.5	0.05	0.38	30
Nitrate N	0.1	2	10	11.3	1
Turbidity	2	20	1	10	–
BOD	2	4	1	–	250
Fluoride	<0.5	<0.5	<1	1.5	–
Iron	0.2	0.2	1	0.2	–
Aluminium	0.1	0.1	0.1	0.2	–
Pesticides	–	–	–	0.0001	–
Phenols	–	–	–	0.0005	–
PAH	–	–	–	0.0002	–
Lead	–	–	–	0.05	–
22 C count/ml	100	30 000	10	–	10
37 C count/ml	10	7 500	5	–	10
Coliform MPN/100 ml	5	1 000	5	<1	10

Values are expressed as mg/l except where noted.
* Maximum Allowable Concentration.

2.4 Water supply and treatment

Most water supplies are obtained from surfacewater sources such as rivers and lakes or from groundwater sources known as aquifers. With many surfacewater sources the intermittent and seasonal nature of precipitation is such that some form of storage must be provided in the catchment to ensure a steady yield. This usually takes the form of an impounding reservoir feeding an aqueduct for a direct supply or providing augmentation of river flow for downstream abstraction in the case of a regulating reservoir. In some parts of the world where rainfall is low, surfacewater sources may be non-existent and recourse has to be made to groundwater, which may be highly saline, or possibly to seawater. In these latter situations the type of treatment necessary to produce an acceptable water supply is costly and quite different from that normally employed for other raw waters. The object of water treatment is to produce a wholesome product which meets the appropriate quality standards. The type of treatment required to attain the desired quality is a function of the raw water quality and, except for very high quality sources, it is often necessary to employ several different treatment processes in series. Water abstracted from lowland river sources normally requires the most comprehensive form of treatment. Such waters are likely to contain significant amounts of fine suspended matter in the form of soil particles, bacteria, etc., as well as soluble inorganic and organic compounds. Treatment therefore usually involves a combination of physical processes such as screening, sedimentation and filtration aided by chemical processes like coagulation and disinfection. Table 2.2 indicates probable treatment systems for a number of raw water sources.

2.5 Wastewater collection, treatment and disposal

In developed countries it is accepted that the most effective way of collecting liquid human and industrial wastes is the water carriage system. This places considerable demands on water resources and is therefore not necessarily appropriate in rural areas or for those in which water is scarce. In temperate urban areas most of the water supplied to homes and industry reappears as wastewater which contains relatively small concentrations of suspended and dissolved organic matter, similar inorganic constituents as the water supplied to the area, large numbers of microorganisms in the case of domestic sewage and possibly toxic materials from certain industrial discharges. A suitable system of sewers, which conveys wastewater and surface run off in the same or different pipes, must be provided to transport the wastewater to a convenient disposal point. The

Table 2.2 Common water treatment systems

Source	*Probable treatment*
Upland catchment	Screening, sand filtration, disinfection, pH adjust Optional depending upon quality: Coagulation or adsorption for removal of colour
Lowland river	Screening, coagulation, sedimentation, filtration, disinfection, pH adjustment Optional depending upon quality: Adsorption for removal of colour and trace organics Softening Aeration
Groundwater	Disinfection Optional depending upon quality: Softening pH adjustment Adsorption for removal of trace organics

degree of treatment required depends upon the location of the discharge and the amount of dilution available but in most countries comprehensive treatment must be provided before discharge to inland waters. Wastewater treatment plants usually employ a series of unit processes to achieve the desired removal of organic matter and suspended material. Physical processes like screening and sedimentation are complemented by biological processes for the stabilization of organic matter. Table 2.3 outlines typical wastewater treatment systems for domestic discharges.

2.6 Pollution control

Pollution of water is undesirable since it can produce potential hazards in drinking water supplies, increase the load on water treatment plants, hinder industrial uses of water, damage fisheries, and cause general deterioration in the amenity and recreational value of the water. Natural watercourses have considerable capacity for self-purification for some pollutants, and this and the downstream use of the receiving water are usually taken into account when assessing allowable levels of pollution in discharges. Many pollution problems are due to the effect on the dissolved oxygen in water, since a balanced aquatic ecosystem requires the presence of sufficient oxygen in the water. Organic pollutants are usually a source of food for

Table 2.3 Common sewage treatment systems

Receiving water	*Probable effluent*		*Probable treatment*
	BOD (mg/l)	*SS*(mg/l)	
Open sea	–	–	Screening
Tidal estuary	150	150	Screening, sedimentation
Lowland river	20	30	Screening, sedimentation, aerobic biological oxidation, sedimentation, anaerobic sludge digestion
Upland river	10	10	As for lowland river plus sand filtration or grass plots
High-quality river with low dilution	5	5	As for upland river plus nutrient removal

microorganisms and in their stabilization oxygen is consumed. Excessive amounts of organic matter discharged to a watercourse will cause depletion of the dissolved oxygen level which, if serious, may kill most forms of aquatic life. Pollution control measures are thus aimed at ensuring satisfactory dissolved oxygen concentrations in the receiving water, removing excessive amounts of suspended matter and preventing the discharge of toxic materials which would prevent full use of the water for other purposes.

FURTHER READING

Henderson-Sellers, B., *Reservoirs*. Macmillan, 1979.

Institution of Water Engineers and Scientists, *Water Supply and Sanitation in Developing Countries*, IWES, 1983.

Price, M., *Introducing Groundwater*, Allen and Unwin, 1985.

Sharp, J. J. and Sawden, P., *BASIC Hydrology*, Butterworths, 1984.

Tebbutt, T. H. Y.,*Water Science and Technology*, John Murray, 1973.

Vallentine, H. R., *Water in the Service of Man*, Penguin, 1967.

Chapter 3

Flow measurement, sampling and analysis

Almost all aspects of water and wastewater treatment require a quantitative understanding of the situation. This means that information must be available about the volume and rate of flow and its quality. Flow measurement, sampling and analysis are thus important factors in the design and operation of treatment systems.

ESSENTIAL THEORY

3.1 Flow measurement techniques

The only direct way of measuring flow is to collect the total discharge in a calibrated container or in a weighing tank over a known period of time. This technique is clearly not very practicable for most situations, so that some indirect means of flow measurement is usually necessary. This will involve the use of a gauging structure, velocity measurements in a known cross-section, or dilution gauging.

3.1.1 Gauging structures

A common and accurate method of flow measurement is to pass the flow through or over a structure which has a known head-discharge relationship. For closed pipes or conduits running under pressure, venturi meters or orifice plates are the normal means of flow measurement. These rely on using the differential pressure across the device to determine the flow. Thus for a venturi meter (Figure 3.1) the discharge is given by

$$Q = Ca[2g(H_1 - H_2)]^{0.5} \tag{3.1}$$

where Q = flow
C = discharge coefficient
a = cross-sectional area of throat
g = acceleration due to gravity

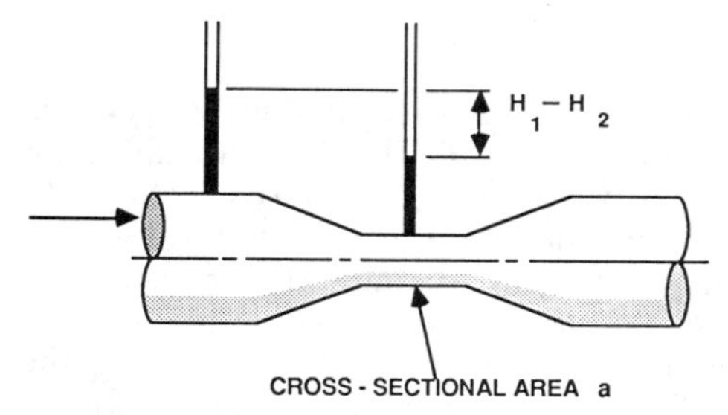

Figure 3.1 Venturi meter

$H_1 - H_2$ = pressure difference between inlet and throat

For open channels, flows can be determined using venturi flumes, sharp-edged rectangular or vee weir plates, broad-crested weirs or triangular profile weirs. The basic equations for thin plate weirs (Figure 3.2), ignoring approach velocity effects, are

Vee notch

$$Q = 0.53 \tan(\theta/2)(2g)^{0.5} H^{2.5} \tag{3.2}$$

Rectangular notch

$$Q = 0.67B(2g)^{0.5} H^{1.5} \tag{3.3}$$

where θ = angle subtended by notch
H = head over notch
B = width of rectangular notch

Measurement of head is usually by means of a float in a stilling well upstream of the structure or with an ultrasonic level detector. Gauging structures can be costly to construct and can cause undesirable restrictions to the flow, so that in many circumstances they may not be appropriate and one of the other flow-gauging techniques must be used.

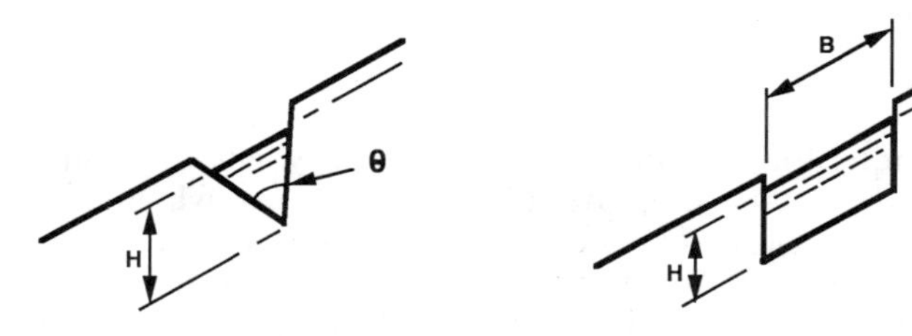

Figure 3.2 Thin plate weirs

3.1.2 Velocity measurements

Since discharge is the product of velocity and cross-sectional area, it is clearly possible to determine discharge in a channel or pipe by measuring the velocity in a known cross-section. In practice the situation is complicated by the fact that velocity is not constant in a flow and in natural channels the cross-section is likely to be irregular. A simple approximation is to take the velocity measured at 0.6 depth from the surface as being the mean velocity of flow. For small channels a float suitably ballasted can be used to give an indication of the 0.6 depth velocity but for more accurate measurements a current meter must be employed. This is a propellor or bucket wheel device which is held in the flow and caused to rotate by movement of water past it. Visual or electronic indications are given after a known number of revolutions and these can be related to velocity by means of a tank calibration. Accuracy of measurements is helped by dividing the section into a number of panels, as shown in Figure 3.3:

$$Q = \Sigma\,[(d_n + d_{n+1})/2][(V_n + V_{n+1})/2]b \qquad (3.4)$$

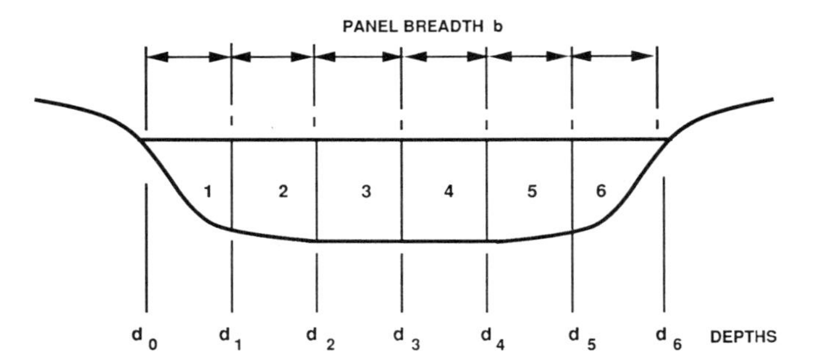

Figure 3.3 Current metering

A newer technique for measuring mean velocity which is becoming increasingly popular employs ultrasonic beams which are transmitted at an angle across the channel and reflected back to a transmitter/receiver. Discontinuities in the flow reflect and attenuate the sound waves so that the pulses travelling in the two directions have slightly different timings due to the velocity of the water.

3.1.3 Dilution gauging

In some situations it is not possible to construct permanent gauging structures, and current metering may be difficult because of access or safety problems. Flow in sewers and effluent discharge culverts are examples of this type of situation where the use of chemical dilution gauging can be advantageous. This technique which can be used in open or closed conduits, involves the addition of a known amount of tracer into the flow and collection of samples at a point downstream where complete mixing of the tracer and the flow has taken place. In the constant rate injection method the tracer is added at a known concentration for sufficient time to reach a constant downstream concentration after mixing has occurred (Figure 3.4). Samples taken from this plateau are analysed to determine the tracer concentration and then using mass balance the flow can be calculated:

$$qc_0 + Qc_1 = (Q + q)c_2 \qquad (3.5)$$

where Q = stream flow
q = injection rate
c_0 = tracer injection in dosing solution
c_1 = background concentration of tracer in stream
c_2 = tracer concentration in stream after mixing

Hence

$$Q = [(c_0 - c_2)/(c_2 - c_1)]q \qquad (3.6)$$

Usually $c_0 \gg c_2$ so that the relationship becomes

$$Q = c_0q/(c_2 - c_1) \qquad (3.7)$$

In selecting a tracer it is necessary to bear in mind the need for the material to have a low and constant natural background, and to be non-toxic, stable, easy to determine, highly soluble, readily available and cheap. There is no universal tracer,

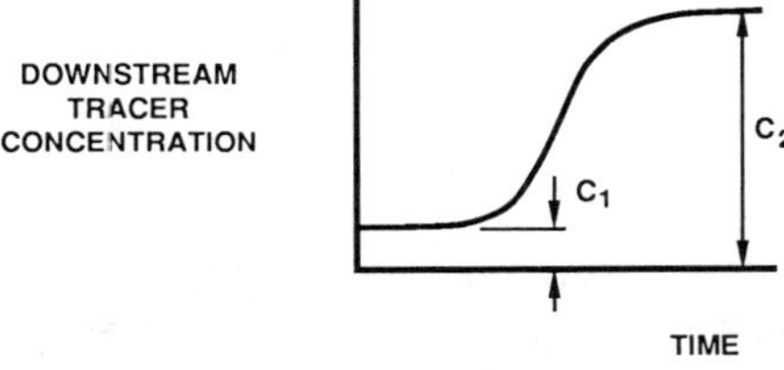

Figure 3.4 Dilution gauging

although common salt fulfils many of the requirements except in sewage or saline waters, where lithium salts or fluorescent dyes are more practical.

3.2 Sampling techniques

Sampling is a very important stage in the assessment of water quality, since if the sample is unrepresentative any analytical results will be useless. Collection of a representative sample from a source of uniform quality is easy and can be achieved by means of a simple grab sample. Such a single sample is also appropriate if the purpose of sampling is to monitor whether or not quality is within a specified limit. With many natural waters and most wastewaters both flow and quality are highly variable, so that a grab sample is unlikely to provide sufficient information to assess pollution loads or to permit evaluation of treatment plant performance. To provide this type of data from a source whose flow and strength varies as shown in Figure 3.5, it is necessary to use a composite sampling technique in which samples are collected manually or automatically at known time intervals when flow measurements are also available. The individual samples are then mixed in proportion to the flow at the time of sampling to produce an integrated sample for analysis. Some quality parameters can be measured continuously but many require laboratory determinations for reliable results to be achieved.

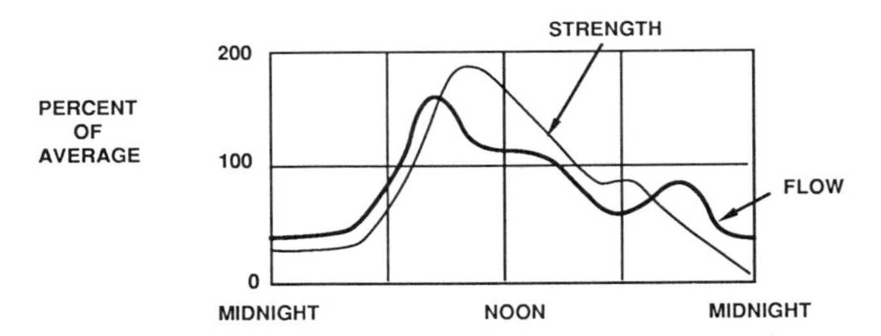

Figure 3.5 Typical diurnal flow and strength curve for sewage

3.3 Analytical methods

The concentrations of impurities found in waters and wastewaters are relatively low compared with normal chemical solutions, so that analytical techniques appropriate to dilute aqueous systems are necessary. In general, concentrations of impurities are expressed in terms of milligrams per litre (mg/l), and analytical procedures must be carefully carried out to obtain accurate information. To permit comparison of data obtained by different analysts it is important to use standard analytical methods such as those published by the Department of the Environment in the UK and the American Public Health Association. There are four main types of physical and chemical analyses which are commonly employed in the examination of waters and wastewaters. These are gravimetric, volumetric, colorimetric, and the use of electrodes. Specialized microbiological analyses are used to detect and enumerate bacteria and other microorganisms in samples.

3.3.1 Gravimetric analysis

This form of analysis relies on weighing solids obtained from a known volume of sample after evaporation, filtration or precipitation. The main uses of gravimetric analysis are found in the determination of total solids by evaporation and of suspended solids by filtration through a 0.45 μm paper. A sensitive analytical balance, drying oven and desiccator are essential for gravimetric determinations, which are not feasible for field use.

3.3.2 Volumetric analysis

Many determinations in water quality can be rapidly and conveniently carried out by volumetric analysis, which depends upon the measurement of volumes of a known strength liquid reagent which reacts with the constituent being determined. The apparatus required is simple and the analyses can usually be carried out in the field if necessary.

3.3.3 Colorimetric analysis

For a wide range of constituents in natural waters and in wastewaters it is possible to utilize the formation of a soluble coloured compound following the addition of a special reagent as a means of analysis. The coloured solution must be such that light absorption through it increases exponentially with the concentration and also that light absorption increases exponentially with the length of the light path through the solution. The colour intensity of a solution can be measured by visual comparison with standards, or more usually by instruments provided with colour filters or prisms to produce the appropriate wavelength of light for greatest sensitivity. Such techniques are widely used for portable analytical systems for field use and also for continuous monitoring of some constituents. Colorimetric analysis is inaccurate in the presence of suspended matter since this also absorbs light and results in a false indication of solution concentration. Nephelometry, which measures the amount of scattered light when a beam is passed through a liquid sample, is used to determine the presence of colloidal solids in a liquid which are responsible for turbidity (a cloudy appearance).

3.3.4 Electrodes

For many years pH ($-\log_{10}[H^+]$), which expresses the intensity of acidity or alkalinity of a solution, has been measured using a glass electrode which is sensitive to hydrogen ions in solution. More recently many other electrodes have become available to measure specific ions such as ammonium, nitrate, chloride, calcium, sodium, etc. Dissolved oxygen is also easily measured by means of a special electrode which is particularly suited for field use. All electrodes require careful, and sometimes frequent, calibration and regular cleaning if left in position for continuous monitoring purposes.

3.3.5 Microbiological analyses

Most determinations in this area are concerned with enumerating bacteria and are based on total viable cell counts on a general-purpose medium or specific medium and incubation conditions for the normal indicator organism *Escherichia coli*, whose presence is taken as positive evidence of human faecal pollution.

3.3.6 Specialized analytical techniques

When analysing samples for specific organic or inorganic constituents, particularly at low concentrations, it is often necessary to employ sophisticated instruments such as atomic absorption spectrophotometers, fluorimeters, gas chromatographs, and mass spectrometers. These instruments are costly to purchase and maintain and require skilled operators to obtain reliable results.

3.4 Some typical analyses

3.4.1 Alkalinity

Alkalinity is due to the presence in water of the anions OH^-, CO_3^{2-} and HCO_3^-, in various combinations. Measurement is by titration with a standard acid whose strength is such that 1 ml is equivalent to 1 mg of $CaCO_3$. Neutralization of OH^- is complete at pH 8.2 whereas neutralization of CO_3^{2-} is only half complete at pH 8.2, full neutralization taking place at pH 4.5. Alkalinity above pH 8.2 is termed caustic to differentiate it from total alkalinity which exists down to pH 4.5. By titrating a sample with acid to the two pH values 8.2 and 4.5, which is undertaken using the appropriate indicators, it is possible to determine the composition of the alkalinity using the following rules.

HCO_3^- and OH^- cannot exist together

OH^- alone gives initial pH > 8.2
OH^- alkalinity = caustic = total

CO_3^{2-} alone gives initial pH > 8.2
CO_3^{2-} alkalinity = 2 × caustic = total

OH^- and CO_3^{2-} together give initial pH > 8.2
CO_3^{2-} alkalinity = 2 × titration from pH 8.2 to 4.5
OH^- alkalinity = total − CO_3^{2-} alkalinity

CO_3^{2-} and HCO_3^- together give initial pH > 8.2
CO_3^{2-} alkalinity = 2 × caustic alkalinity
HCO_3^{2-} alkalinity = total − CO_3^{2-} alkalinity

HCO_3^- alone gives pH < 8.2
HCO_3^- alkalinity = total alkalinity

3.4.2 Biochemical oxygen demand (BOD)

When organic compounds are oxidized by microorganisms in the presence of oxygen, the rate at which oxygen is utilized can be used as a measure of the strength and biodegradability of the organic matter. Exertion of BOD is assumed to be a first-order reaction such that

$$dL/dt = -KL \quad (3.8)$$

where L = organic matter remaining to be oxidized
t = time
K = rate constant

Integrating

$$L_t/L = e^{-Kt} \tag{3.9}$$

where L_t = BOD remaining at time t

The usual concern is for BOD already exerted, which is given by

$$BOD_t = (L - L_t) = L(1 - e^{-Kt}) \tag{3.10}$$

Because the BOD reaction is biochemical it proceeds relatively slowly and thus the standard incubation conditions of 5 days at 20°C will result in oxidation of about 65% of the biodegradable matter in sewage. To obtain the values of K and L for a particular sample, it is necessary to carry out daily BOD determinations over a period of 7–10 days. The variability of biological growth means that the data do not exactly fit the model of Equation 3.10 and some form of curve fitting must be employed to evaluate the best values of K and L. A convenient computational method is that due to Thomas, which relies on the similarity of the expansions of the two expressions

$$(1 - e^{-Kt}) \text{ and } Kt(1 + Kt/6)^{-3}$$

Thus Equation 3.10 can be approximated by

$$BOD_t = LKt(1 + Kt/6)^{-3} \tag{3.11}$$

i.e.

$$(t/BOD_t)^{1/3} = (KL)^{-1/3} + (K^{2/3}/6L^{1/3})t \tag{3.12}$$

This gives a straight line with intercept $(I) = (KL)^{-1/3}$ and slope $(S) = K^{2/3}/6L^{1/3}$. Hence $K = 6S/I$ and $L = 1/KI^3$.

The experimental results can thus be plotted and a visual line of best fit drawn or a computed line can be used to obtain the values of K and L.

3.5 Interpretation of results

When utilizing field or laboratory data it is important to appreciate the degree of accuracy and the natural variation in measurements. It is important not to impute a greater degree of accuracy to measurements than is warranted by the actual data. The reporting of results to eight decimal places from a calculator display or computer printout is meaningless if one of the terms used in obtaining the results is only correct to one decimal place. The concept of significant figures is important to the credibility of reported results, which should not be quoted to a level of accuracy higher than that of the least accurate measurement in a system.

The subject of statistical analysis of data is well covered in many excellent texts and it is impossible to even attempt to provide an introduction to the topic in a few paragraphs. Nevertheless there are a number of frequently used statistical tools which should be mentioned here. Standard deviation is the primary measure of variation in a number of measurements and it is obtained by calculating the mean value, summing the squares of the differences between individual values and the mean, dividing by the number of measurements and taking the square root. Thus for n measurements, the standard deviation σ is given by:

$$\sigma = \sqrt{\left[\frac{\Sigma(x - \bar{x})^2}{n - 1}\right]} \qquad (3.13)$$

where x = an individual measurement
$\bar{x}$ = mean of n measurements

For a normal probability distribution it is useful to remember that 62.86% of values fall within $\bar{x} \pm \sigma$, and 95.46% of values fall within $\bar{x} \pm 2\sigma$. This information can often be used to give an approximate indication of the reliability of results and also of whether or not two sets of data are likely to be from different populations. There are, of course, much more specific statistical analyses for testing for significant differences between data sets.

Another area of data analysis which is of considerable value is that of regression to determine relationships between variables, and the closely allied property correlation, which indicates the degree of interdependence between variables. In a simple linear regression the variables are related by means of the equation

$$y = mx + c \qquad (3.14)$$

where m = slope of line
c = intercept on y axis

The intercept c is obtained by summing the product of the individual deviations of the x and y values from their respective means and dividing the results by the sum of the squares of the individual x deviations from the mean:

$$c = \sum [(x - \bar{x})(y - \bar{y})]/\sum(x-\bar{x})^2 \qquad (3.15)$$

The correlation coefficient is obtained from the expression

$$r_{xy} = \sum[(x - \bar{x})(y - \bar{y})]/\sqrt{[\sum(x - \bar{x})^2\sum(y - \bar{y})^2]} \qquad (3.16)$$

A perfect relationship between two variables gives a correlation coefficient of 1 and the absence of any relationship gives a correlation coefficient of 0. The reliability of the correlation coefficient is inversely related to the square root of the number of measurements, so that its significance depends upon the number of measurements, and for small numbers of measurements it is common to obtain correlation coefficients close to unity with little practical significance. A correlation coefficient of, say, 0.9 is not necessarily very meaningful if based on only a few measurements but would be highly significant if obtained from a large database.

WORKED EXAMPLES

Example 3.1 CURMET: current metering

A program is required to take the depth and mean velocity readings from a river gauging and use the data to calculate the mean discharge of the river. The program should be capable of dealing with different numbers of panels and different panel widths. The user should be prompted for the appropriate depth and velocity values at the edges of each panel after specifying the number of panels used for the gauging.

A current meter gauging of a river 11 m wide was undertaken by making velocity measurements at 0.6 depth from the surface at 10 equally spaced positions across the section, with the results given in the following table:

Position number	*Depth* (m)	*Velocity* (m/s)
0	0	0
1	1.5	1.1
2	1.8	2.0
3	2.0	3.4
4	2.1	4.8
5	2.4	5.1
6	2.7	4.9
7	2.5	4.4
8	2.1	3.1
9	1.5	1.9
10	1.0	1.0
11	0	0

Determine the discharge.

```
10 CLS
20 REM CURMET
30 PRINT"Current Metering Calculations"
40 INPUT"Enter number of panels, panel width (m)";N,B
50 FOR P=1 TO N
60 PRINT"Data for Panel";P
70 INPUT"Enter depths (m) at panel edges (use comma)";D1,D2
80 INPUT"Enter left, right mean velocities (m/s)";V1,V2
90 QP=(.5*(D1+D2)*.5*(V1+V2))*B
100 QT=QT+QP
110 NEXT P
120 PRINT"Discharge in section=";QT;"cu.m/s"
130 PRINT
140 INPUT"Another set of results (Y/N)";Q$
150 IF Q$="Y" OR Q$="y" THEN 10
160 IF Q$ <> "N" OR Q$ <> "n" THEN 170
170 END
180 PRINT"Enter Y or N only ....you entered ";Q$
190 GOTO 130
200 END
```

```
Enter number of panels, panel width (m)? 11,1
Data for Panel 1
Enter depths (m) at panel edges (use comma)? 0,1.5
Enter left, right mean velocities (m/s)? 0,1.1
Data for Panel 2
Enter depths (m) at panel edges (use comma)? 1.5,1.8
Enter left, right mean velocities (m/s)? 1.1,2
Data for Panel 3
Enter depths (m) at panel edges (use comma)? 1.8,2
Enter left, right mean velocities (m/s)? 2,3.4
Data for Panel 4
Enter depths (m) at panel edges (use comma)? 2,2.1
Enter left, right mean velocities (m/s)? 3.4,4.8
Data for Panel 5
Enter depths (m) at panel edges (use comma)? 2.1,2.4
Enter left, right mean velocities (m/s)? 4.8,5.1
Data for Panel 6
Enter depths (m) at panel edges (use comma)? 2.4,2.7
Enter left, right mean velocities (m/s)? 5.1,4.9
Data for Panel 7
Enter depths (m) at panel edges (use comma)? 2.7,2.5
Enter left, right mean velocities (m/s)? 4.9,4.4
Data for Panel 8
Enter depths (m) at panel edges (use comma)? 2.5,2.1
Enter left, right mean velocities (m/s)? 4.4,3.1
Data for Panel 9
Enter depths (m) at panel edges (use comma)? 2.1,1.5
Enter left, right mean velocities (m/s)? 3.1,1.9
Data for Panel 10
Enter depths (m) at panel edges (use comma)? 1.5,1
Enter left, right mean velocities (m/s)? 1.9,1
Data for Panel 11
Enter depths (m) at panel edges (use comma)? 1,0
Enter left, right mean velocities (m/s)? 1,0
Discharge in section= 67.67 cu.m/s

Another set of results(Y/N)? N
```

Program notes

(1) Lines 10–30 set up title
(2) Line 40 asks for details about section
(3) Line 50 sets up loop for data input
(4) Lines 60–80 ask for panel data
(5) Line 90 calculates discharge in panel
(6) Line 100 accumulates discharge in panels
(7) Line 110 returns loop for next panel
(8) Line 120 prints total discharge in section
(9) Lines 130–140 offer another calculation if desired
(10) Line 150 traps error in incorrect response to 130
(11) Line 160 ends program if no further calculation
(12) Line 170 error message for incorrect response to 130
(13) Line 180 redirects program after incorrect response

Example 3.2 CHEMDIL: chemical dilution gauging

Write a program which will produce a table of tracer concentrations and corresponding flows for use in dilution flow gauging. The user should be able to input all relevant parameters and specify the range of downstream tracer concentrations to be covered.

Flow measurement in a trunk sewer was carried out using lithium nitrate as the tracer. The lithium concentration in the tracer solution was 12.5 g/l and the injection rate was 5 ml/s. The background concentration of lithium in the sewer was 0.05 mg/l. Produce a table of sewer flows corresponding to downstream tracer concentrations in the range 0.5–2.5 mg/l.

```
10 CLS
20 REM CHEMDIL
30 PRINT"Chemical Dilution Gauging Results Table"
40 INPUT"Enter tracer background concn (mg/l)";C1
50 INPUT"Enter tracer flow rate (ml/s), concn (g/l)";Q0,C0
60 PRINT"Now enter minimum and maximum expected values of"
70 PRINT"downstream tracer concns (mg/l)"
80 INPUT CL,CU
90 INPUT"Enter concn increment for step calcn (mg/l)";S
100 CLS
110 PRINT"Tracer Concn","Discharge"
120 PRINT"  (mg/l)    ","   (l/s) "
130 FOR C2=CL TO CU STEP S
140 Q=C0*Q0/(C2-C1)
150 PRINT C2,Q
160 NEXT C2
170 END
```

```
Chemical Dilution Gauging Results Table
Enter tracer background concn (mg/l)? .05
```

```
Enter tracer flow rate (ml/s), concn (g/l)? 5,12.5
Now enter minimum and maximum expected values of
downstream tracer concns (mg/l)
? .5,2.5
Enter concn increment for step calcn (mg/l)? .25

Tracer Concn   Discharge
  (mg/l)          (l/s)
 .5             138.8889
 .75            89.28571
 1              65.78948
 1.25           52.08333
 1.5            43.10345
 1.75           36.76471
 2              32.05128
 2.25           28.40909
 2.5            25.51021
0
```

Program notes

(1) Lines 10–30 set up title
(2) Lines 40–50 ask for basic parameters for gauging
(3) Lines 60–90 input specification for output table
(4) Line 100 clears screen for output
(5) Lines 110–120 print heading for output table
(6) Line 130 sets up loop for specified limits and increment
(7) Line 140 calculates flow [(g/l × ml/s)/(mg/l) = l/s]
(8) Line 150 prints results
(9) Line 160 returns loop for next value of concentration

Example 3.3 ALKTY: alkalinity analysis

A program is required which takes the results from alkalinity titrations and calculates total and caustic alkalinity and also identifies the anions causing the alkalinity. It should allow for different sizes of samples in the titration.

An example of results from an alkalinity titration is:

sample size: 25.0 ml
titration to pH 8.2: 5.8 ml
total titration to pH 4.5: 18.2 ml

```
10 CLS
20 REM ALKTY
30 PRINT"Alkalinity Determinations"
40 PRINT"Titration for Total and Caustic Alkalinity"
50 PRINT
60 INPUT"Enter sample volume (ml)";S
70 INPUT"Enter vol of acid to phenolphthalein end point (ml)";C
80 INPUT"Enter vol of acid to methyl orange end point (ml)";T
90 IF T<C THEN PRINT"***ERROR IN DATA***":GOTO 70
100 CA=C*1000/S
110 TA=T*1000/S
120 PRINT"Total alkalinity=";TA;"mg/l as calcium carbonate"
```

```
130 PRINT"Caustic alkalinity=";CA;"mg/l as calcium carbonate"
140 PRINT
150 IF CA=0 THEN GOTO 230
160 IF CA<.5*TA THEN 210
170 PRINT"Hydroxide alkalinity=";TA-2*(TA-CA);"mg/l"
180 IF CA=TA THEN 240
190 PRINT"Carbonate alkalinity=";2*(TA-CA);"mg/l"
200 GOTO 220
210 PRINT"Carbonate alkalinity=";2*CA;"mg/l"
220 IF TA<=2*CA THEN 240
230 PRINT"Bicarbonate alkalinity=";TA-2*CA;"mg/l"
240 PRINT
250 INPUT"Another set of results (Y/N)";Q$
260 IF Q$="Y" OR Q$="y" THEN 60
270 END
```

```
Alkalinity Determinations
Titration for Total and Caustic Alkalinity

Enter sample volume (ml)? 25
Enter vol of acid to phenolphthalein end point (ml)? 5.8
Enter vol of acid to methyl orange end point (ml)? 18.2
Total alkalinity= 728 mg/l as calcium carbonate
Caustic alkalinity= 232 mg/l as calcium carbonate

Carbonate alkalinity= 464 mg/l
Bicarbonate alkalinity= 264 mg/l

Another set of results (Y/N)? Y
Enter sample volume (ml)? 25
Enter vol of acid to phenolphthalein end point (ml)? 9.2
Enter vol of acid to methyl orange end point (ml)? 14.9
Total alkalinity= 596 mg/l as calcium carbonate
Caustic alkalinity= 368 mg/l as calcium carbonate

Hydroxide alkalinity= 140 mg/l
Carbonate alkalinity= 456 mg/l

Another set of results (Y/N)? N
```

Program notes

(1) Lines 10–50 set up title
(2) Lines 60–80 input titration data
(3) Line 90 checks for incorrect data, total alkalinity cannot be less than the caustic alkalinity, returns to 70 if error exists
(4) Lines 100–140 calculate and print alkalinity results
(5) Line 150 condition for bicarbonate alkalinity only, if true, program jumps to 230
(6) Line 160 condition for bicarbonate and carbonate alkalinity, if true, program jumps to 210
(7) Line 170 calculates and prints hydroxide alkalinity
(8) Line 180 condition for hydroxide alkalinity only, if true, program jumps to 240
(9) Line 190 calculates and prints carbonate alkalinity
(10) Line 200 causes program to skip lines which refer to forms of alkalinity which have been eliminated by this stage

(11) Line 210 calculates and prints carbonate alkalinity
(12) Line 220 condition for no bicarbonate alkalinity, if true, program jumps to 240
(13) Line 230 calculates and prints bicarbonate alkalinity
(14) Lines 240–260 offer another calculation if desired

Example 3.4 BODDATA: BOD data analysis

Write a program to accept BOD data obtained over several days and use the Thomas Method to calculate the ultimate BOD and the rate constant for the sample.

An example of such data is:

Day	1	2	3	4	5	6	7
BOD	38	58	76	94	98	104	112

```
10 CLS
20 REM BODDATA
30 PRINT"BOD Data Analysis (Thomas Method)"
40 X=0:Y=0:X2=0:Y2=0:XY=0
50 INPUT"Enter number of observations";N
60 FOR D=1 TO N
70 INPUT"Enter time (d), and BOD (mg/l)";T,B
80 P=(T/B)^(1/3)
90 X=X+T
100 Y=Y+P
110 X2=X2+T^2
120 Y2=Y2+P^2
130 XY=XY+T*P
140 NEXT D
150 S=(N*XY-Y*X)/(N*X2-X^2)
160 I=(Y-S*X)/N
170 K=6*S/I
180 L=1/(K*I^3)
190 K1=.4343*K
200 CLS
210 PRINT"Ultimate BOD=";L;"mg/l"
220 PRINT"Rate constant (base 10)=";K1;"per day"
230 IF B>.9*L THEN 280
240 PRINT
250 INPUT"Another set of results (Y/N)";Q$
260 IF Q$="Y" OR Q$="y"THEN 40
270 END
280 PRINT"Final BOD value too close to L value"
290 PRINT"Re-enter data omitting final BOD value"
300 GOTO 40
```

```
BOD Data Analysis (Thomas Method)
Enter number of observations? 7
Enter time (d), and BOD (mg/l)? 1,38
Enter time (d), and BOD (mg/l)? 2,58
Enter time (d), and BOD (mg/l)? 3,76
Enter time (d), and BOD (mg/l)? 4,94
```

```
Enter time (d), and BOD (mg/l)? 5,98
Enter time (d), and BOD (mg/l)? 6,104
Enter time (d), and BOD (mg/l)? 7,112

Ultimate BOD= 124.8741 mg/l
Rate constant (base 10)= .1455418 per day

Another set of results (Y/N)?
```

Program notes

(1) Lines 10–30 set up title
(2) Line 40 initializes variables for repeated runs
(3) Lines 50–70 set up loop to input daily data
(4) Lines 80–130 least squares calculations
(5) Line 140 returns loop for next input
(6) Lines 150–160 calculate slope and intercept for straight line fit to data
(7) Line 170 calculates rate constant
(8) Line 180 calculates ultimate BOD
(9) Line 190 converts rate constant to base 10
(10) Lines 200–220 print results
(11) Line 230 checks validity of data, since method becomes inaccurate if last BOD is close to calculated ultimate BOD; if true, program jumps to 270 for warning
(12) Lines 240–250 offer another calculation if desired

The program can be modified to use results given in a data format by making the following changes:

```
70  READ T, B
240 INPUT"Change data set (Y/N)";Q$
250 IF Q$="Y" OR Q$="y" THEN 300
290 GOTO 300
300 DATA 1,38,2,58,3,76,4,94,5,98,6,104,7,112
```

PROBLEMS

(3.1) Write a program which produces a table of flow against head for a rectangular notch using Equation 3.3. Use the program to give flow data for a notch 0.55 m wide and for a range of heads of 5 to 200 mm.

(3.2) A rectangular notch 350 mm wide is used to gauge flow in a sewer at a point where a sampler collects individual samples at hourly intervals over a period of 24 hours. A float gauge records level measurements at the time of sample collection. Develop the program from the problem above to calculate the appropriate volumes of the individual hourly samples to be combined to give a 2-litre composite sample of the flow over the 24-hour period.

(3.3) The determination of suspended solids involves filtration of a known volume of liquid through a preweighed filter paper and drying to constant weight at 103°C. The increase in weight gives the total suspended solids (SS). By firing at 600°C, organic matter is removed so that volatile suspended solids (VSS) are measured by the loss on firing. Write a program which will calculate total and volatile SS values expressed in mg/l when the sample volume is in ml and the weights in g . Typical results for a 50 ml sample of settled sewage would be:

initial weight of paper: 0.2016 g
weight after drying: 0.2091 g
weight after firing: 0.2076 g

(3.4) Using Equation 3.1, write a program which will tabulate flows through a venturi meter for a given range of pressure differences. The user will be asked to specify the discharge coefficient and the throat area.

FURTHER READING

Department of the Environment, *Methods for the Examination of Waters and Associated Materials*, HMSO. [Separate publications for individual parameters with various dates of publication.]

American Public Health Association, *Standard Methods for the Examination of Water and Wastewater*, APHA, 1985.

Hutton, L. G., *Field Testing of Water in Developing Countries*, WRc, 1983.

Sawyer, C. N. and McCarty, P. L., *Chemistry for Sanitary Engineers*, 2nd edn, McGraw Hill, 1978.

Smith, P. D., *BASIC Hydraulics*, Butterworths, 1982.

Tennant-Smith, J., *BASIC Statistics*, Butterworths, 1984.

Chapter 4

Environmental aspects

In most, if not all, developed countries the last twenty years or so have seen a considerable growth in public interest in environmental matters. This has had a significant effect on the way in which water and wastewater operations are undertaken and the emphasis given to various aspects of pollution control.

It is important to appreciate that, strictly speaking, pure water does not exist in nature. Even rainwater contains a small amount of impurities, largely in the form of inorganic salts derived from natural sources, although supplemented by other constituents arising from the combustion of fossil fuels, the use of aerosols, etc. In relation to water pollution it is helpful if the term is used in relation to situations where the materials added to the water may restrict the use of that receiving water for particular purposes. Thus water pollution is undesirable for many reasons, including:

1. Contamination of water sources resulting in potential disease hazards and increased demand on water treatment.
2. Effects on fish and other aquatic life.
3. Health hazards of irrigation with polluted water.
4. Restrictions on amenity and recreational uses.
5. Creation of odour nuisances.
6. Hindrance to navigation by sludge deposits.

To assess the importance of water pollution in a particular case it is essential to have some way of determining the use to which the receiving water will be put. A degree of pollution which would be quite unacceptable in a river used for salmon fishing and drinking water abstraction could be insignificant in the lower reaches of a river used mainly for drainage and navigation. The use of fixed emission standards to control pollution does not allow for different usages, and although popular with bureaucrats, such pollution control measures are not scientifically valid. A more logical approach which is favoured in the UK is to adopt river quality objectives which take into account the use of the receiving water and the dilution available. A typical water use classification, in order of decreasing water quality requirements, would be:

1. Domestic water supply.
2. Game fishing, both commercial and sport.
3. Industrial water supply.
4. Irrigation.
5. Recreation and amenity.
6. Navigation.
7. Drainage and waste disposal.

A natural watercourse in an unpolluted state has a balanced composition and supports a varied population of living organisms. This balance is largely dependent upon the presence of sufficient amounts of dissolved oxygen and the absence of toxic materials. Thus in developed countries concern about water pollution is mainly directed towards the effect of pollution on the oxygen balance and the possible toxic effects of discharges. In developing countries, where much of the drinking water is supplied without effective treatment, the primary concern in relation to pollution must be with the dangers arising from the entry of pathogenic microorganisms to the environment and the consequent hazards of water-related diseases.

ESSENTIAL THEORY

4.1 Mass balance concepts

It is obviously important to be able to assess the effect of a particular pollutant discharge on a receiving water in quantitative terms, and a mass balance analysis, as used in chemical dilution gauging (Section 3.1), is a fundamental tool in this respect.

Figure 4.1 shows a river with an effluent discharge for which, by assuming instantaneous mixing and conservation of mass, it is possible to quantify the immediate downstream consequences of the discharge.

$$Q_1c_1 + Q_2c_2 = Q_3c_3 \tag{4.1}$$

Since $Q_1 + Q_2 = Q_3$ it is possible to determine the downstream concentration c_3. This concentration will clearly be a function of the dilution available for the effluent and the concentration of the pollutant concerned. Depending upon the nature of the pollutant and the use of the receiving water, a decision can then be made on the likely consequences of the pollution and the need for remedial or pollution control measures.

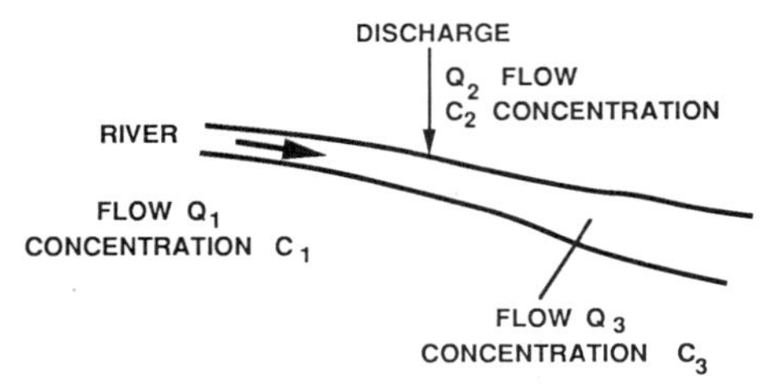

Figure 4.1 Mass balance concept

4.2 Types of pollutant

Pollutants behave in a variety of ways when added to water and this must be taken into account when considering the effects of and control of pollution. Most organic compounds and micro-organisms and some inorganic compounds are degraded by natural processes of self-purification, so that their concentrations decrease with time. This type of pollutant is termed non-conservative, and the actual rate of decay is a function of many factors, including the particular pollutant, the quality of the receiving water, temperature and other environmental factors. For many non-conservative pollutants the decay is reasonably closely modelled as an exponential function:

$$c_t = c_0 e^{-Kt} \tag{4.2}$$

where c_t = pollutant concentration at time t
c_0 = pollutant concentration just below discharge at $t = 0$
K = decay constant for that particular pollutant and temperature

Many inorganic substances are unaffected by self-purification reactions, so that the concentration of pollutant below the discharge point does not change with time unless further dilution water is added in the form of tributary flows with a lower concentration of the pollutant. Such pollutants are termed conservative, and since they are often unaffected by normal water treatment processes, their presence in water may limit its use for certain purposes.

In most situations the following characteristics of pollutants are important:

1. Materials which affect the oxygen balance of the receiving water: substances which consume oxygen, substances which hinder oxygen transfer, thermal pollutants which alter the solubility of oxygen.
2. Toxic substances which inhibit or stop biological activity in the receiving water.
3. High concentrations of inert suspended or dissolved solids which can damage aquatic life.

The effect of pollutants on the oxygen balance of a water is often the major factor in pollution, particularly when the pollutants are organic in nature such as domestic sewage or industrial wastes from food-processing operations. Most of the stabilization processes which occur with such pollutants are biochemical in nature and depend upon the activities of micro-organisms. Organic matter is utilized as a source of food and energy by microorganisms but the way in which this occurs depends upon the availability of oxygen in the environment. There are two basic types of reaction, which are pictured in Figure 4.2. The aerobic reaction takes place in the presence of free oxygen and proceeds fairly rapidly with the release of considerable amounts of energy and leaves stable, relatively inert end products. In the absence of oxygen a two-stage reaction takes place with a much lower energy release and the

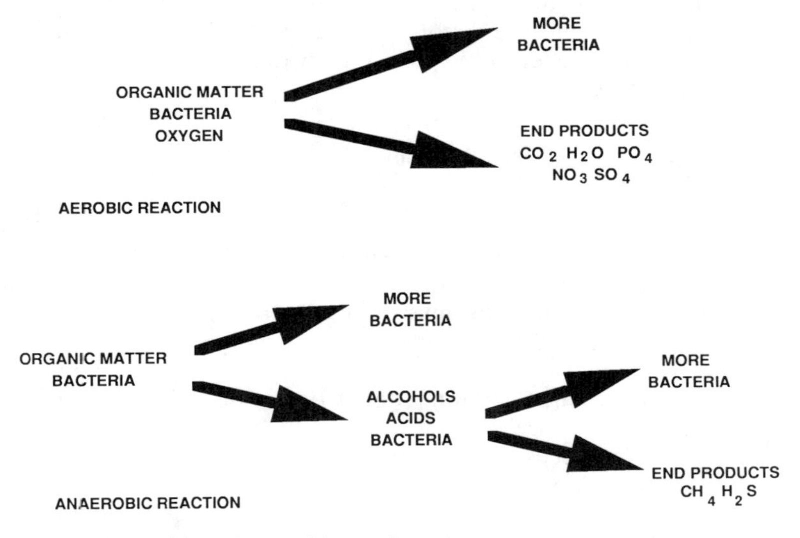

Figure 4.2 Aerobic and anaerobic reactions

formation of relatively unstable end products with a high energy content.

In most circumstances the aerobic reaction is much to be preferred because of its speed and the fact that the continued existence of dissolved oxygen is essential for the higher forms of aquatic life. With an anaerobic reaction the absence of dissolved oxygen means that most forms of aquatic life will die and the water will become objectionable in appearance and smell. Unfortunately, oxygen is only slightly soluble in water (9.2 mg/l at 20°C) and the presence of even quite small amounts of organic matter in a water can be sufficient to quickly cause depletion of the dissolved oxygen levels. This will lead to fish kills and general deterioration in water quality. A primary requirement in pollution control and the design of wastewater treatment facilities is therefore to maintain satisfactory dissolved oxygen levels in the receiving water.

4.3 Reaeration and the oxygen balance

When dissolved oxygen (DO) in a water falls below the saturation level because of some chemical or biochemical activity within the body of water, there will be a transfer of oxygen from the air above the water surface through the air–water boundary. The solubility of a gas in a liquid is a function of the partial pressure of the gas in the atmosphere above the liquid, the temperature, and the dissolved solids content of the liquid.

The rate of transfer of oxygen into water is proportional to the saturation deficit, i.e.

$$\mathrm{d}D/\mathrm{d}t = -KD \tag{4.3}$$

which by integration gives

$$D_t = D_a \mathrm{e}^{-Kt} \tag{4.4}$$

where D_t = DO deficit at time t (saturation DO − actual DO)
D_a = DO deficit at time 0
K = reaeration constant

An alternative arrangement of this expression is

$$\log_e[(c_s - c_t)/(c_s - c_0)] = -Kt \tag{4.5}$$

where c_s = saturation DO
c_0 = initial DO
c_t = DO at time t

The reaeration coefficient K is a function of the velocity of flow, channel configuration, and temperature, and varies widely. For a stagnant pool of water, reaeration takes place entirely by diffusion, which is very slow, but with a rapidly flowing stream, turbulent mixing greatly accelerates oxygen transfer. When considering the discharge of oxygen-consuming pollutants it is therefore important to take into account the oxygen transfer characteristics of the receiving water. The reaeration coefficient can be determined experimentally by artificial depletion of the oxygen levels in a stretch of river followed by monitoring of the recovery. Alternatively, a number of empirical relationships are available which predict reaeration coefficient values on the basis of mean velocity and mean depth of flow.

Using mathematical expressions for BOD uptake and oxygen transfer it is possible to formulate a mathematical model for the dissolved oxygen concentration downstream of a discharge or confluence. The model simplifies the situation in that only the processes of BOD uptake and reaeration are taken into account, but since these are normally by far the most important factors the simplification does not usually produce significant errors.

If the water is originally saturated with DO, the BOD uptake curve for the mixture of stream water and pollutant discharge gives the cumulative deoxygenation of the flow. As soon as the DO begins to fall, an oxygen deficit is created so that reaeration commences. With increasing BOD uptake and increasing DO deficit the rate of reaeration increases in response until a point is reached at which the oxygen enters the water more rapidly than it is being consumed. The oxygen deficit then begins to decrease and, given sufficient time, will return to zero again. The situation is pictured in Figure 4.3 and can be expressed mathematically as

$$\mathrm{d}D/\mathrm{d}t = K_1 L - K_2 D_t \tag{4.6}$$

where D_t = DO deficit at time t
L = ultimate BOD
K_1 = BOD rate constant
K_2 = reaeration constant

Integrating and changing to base 10 ($k = 0.4343K$):

$$D_t = [k_1 L_a/(k_2 - k_1)](10^{-k_1 t} - 10^{-k_2 t}) + D_a 10^{-k_2 t} \tag{4.7}$$

where D_a and L_a are DO deficit and ultimate BOD respectively just below the point of mixing.

The critical point at which the maximum deficit is reached is given by

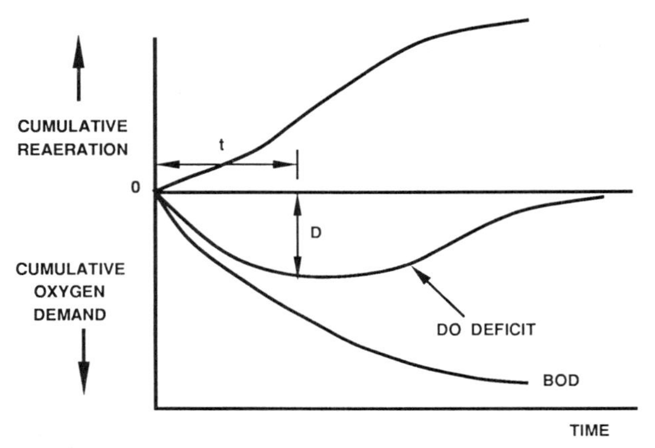

Figure 4.3 The dissolved oxygen balance in a stream

$$dD/dt = 0 = k_1 L - k_2 D_c \tag{4.8}$$

It can be shown that

$$D_c = (k_1/k_2)L_a 10^{-k_1 t_c} \tag{4.9}$$

$$t_c = [1/(k_2 - k_1)] \log \{k_2/k_1[1 - D_a(k_2 - k_1)/L_a k_1]\} \tag{4.10}$$

where D_c = critical deficit at time t_c.

To use the above model, known as the Streeter Phelps equation, it is necessary to know the BOD rate constant and the reaeration constant for the particular stretch of the river under consideration. If the BOD and DO levels upstream of the discharge are known together with the equivalent values for the discharge, mass balances give the BOD and DO immediately downstream of the discharge point. It must be remembered that BOD values are normally given as five-day values which must be converted to ultimate values for incorporation into the Streeter Phelps equation using the following relationship:

$$L = BOD_5/(1 - 10^{-k_1 t}) \tag{4.11}$$

When considering an actual river system in which there may be tributaries and other discharges, the simple model described above can be used to provide a stepwise solution by dividing up the river into a number of reaches. Mass balances at the start of a reach provide the inputs to the model, and at the end of the reach the predicted DO and the remaining BOD, calculated from an exponential decay function, provide the inputs to the following reach.

WORKED EXAMPLES

Example 4.1 BALDEC: pollutant concentration and decay calculations

Write a program to carry out a mass balance calculation for the concentration of pollutants below the confluence of two flows. The user is to have the option of calculating the decay of pollutant for specified distances downstream of the mixing point.

A stream with BOD 2 mg/l has a flow of 2.26 m^3/s and receives a wastewater discharge of 0.775 m^3/s with BOD 30 mg/l. Calculate the BOD concentration just below the discharge point and at a point 3 km downstream if the BOD rate constant k is 0.15 day^{-1} and the velocity of flow is 0.20 m/s.

```
10 CLS
20 REM BALDEC
30 PRINT"Pollutant Concentration and Decay Calculations"
40 INPUT"Enter first, and second flows";F1,F2
50 INPUT"Enter 1st, 2nd pollutant concns (mg/l)";C1,C2
60 C3=(F1*C1+F2*C2)/(F1+F2)
70 PRINT"Pollutant concn just below discharge=";C3;"mg/l"
80 PRINT
90 INPUT"Would you like decay calculations (Y/N)";Q$
100 IF Q$="Y" OR Q$="y" THEN 120
110 END
120 INPUT"Enter velocity of flow below discharge (m/s)";V
130 INPUT"Enter distance downstream for decay calc (km)";D
140 INPUT"Enter decay constant (base 10 per day)";K1
150 T= D/(60*60*24*.001*V)
160 CT=C3*EXP(-K1*2.3026*T)
170 PRINT"Pollutant concn at";D;"km downstream=";CT;"mg/l"
180 PRINT
190 INPUT"Another decay calculation (Y/N)";Q$
200 IF Q$="Y" OR Q$="y" THEN 120
210 END
```

```
Pollutant Concentration and Decay Calculations
Enter first, and second flows? 2.26,0.775
Enter 1st, 2nd pollutant concns (mg/l)? 2,30
Pollutant concn just below discharge= 9.149919 mg/l

Would you like decay calculations (Y/N)? Y
Enter velocity of flow below discharge (m/s)? .2
Enter distance downstream for decay calc (km)? 3
Enter decay constant (base 10 per day)? .15
Pollutant concn at 3 km downstream= 8.617383 mg/l

Another decay calculation (Y/N)?
```

Program notes

(1) Lines 10–30 set up title
(2) Lines 40–50 enter flows and concentrations

(3) Line 60 calculates downstream concentration
(4) Line 70 prints downstream concentration
(5) Lines 80–100 option for decay calculation
(6) Lines 120–140 enter data for decay calculation
(7) Line 150 calculates time
(8) Line 160 calculates pollutant decay
(9) Line 170 prints result
(10) Lines 180–200 offer another calculation if desired

Example 4.2 REAER: reaeration calculations

Write a program to take data from a deoxygenation study on a stream and calculate the reaeration coefficient. The user should then be able to use the calculated coefficient to predict conditions when the initial dissolved oxygen in the reach is different. reach is different.

The following data were obtained during an experimental study of deoxygenation: saturation DO 9.1 mg/l, initial DO 5.6 mg/l, final DO 8.5 mg/l after 4 hours.

```
10 CLS
20 REM REAER
30 PRINT"Reaeration Calculations"
40 INPUT"Enter saturation DO for test conditions (mg/l)";DS
50 INPUT"Enter initial, and final DO levels (mg/l)";DI,DT
60 INPUT"Enter time interval between DO observns (h)";TH
70 K=-LOG((DS-DT)/(DS-DI))*24/TH
80 K2=K*.4343
90 PRINT"Reaeration coefficient (base 10 per day)=";K2
100 PRINT
110 INPUT"Calculations for other conditions/times (Y/N)";Q$
120 IF Q$="Y" OR Q$="y" THEN 140
130 END
140 INPUT"Enter new initial DO (mg/l)";DI
150 PRINT
160 PRINT"Time (d)","DO (mg/l)"
170 FOR T=.25 TO 2 STEP .25
180 DT=(DS-DI)*EXP(-K*T)
190 D=DS-DT
200 PRINT T,D
210 NEXT T
220 PRINT
230 INPUT"Another set of DO predictions (Y/N)";Q$
240 IF Q$="Y" OR Q$="y" THEN 140
250 PRINT
260 INPUT"Another set of calculations (Y/N)";Q$
270 IF Q$="Y" OR Q$="y" THEN 10
280 END
```

```
Reaeration Calculations
Enter saturation DO for test conditions (mg/l)? 9.1
Enter initial, and final DO levels (mg/l)? 5.6,8.5
Enter time interval between DO observns (h)? 4
Reaeration coefficient (base 10 per day)= 4.595558
```

```
Calculations for other conditions/times (Y/N)? y
Enter new initial DO (mg/l)? 6.3

Time (d)        DO (mg/l)
 .25             8.901261
 .5              9.085894
 .75             9.098999
 1               9.099929
 1.25            9.099994
 1.5             9.100001
 1.75            9.100001
 2               9.100001

Another set of DO predictions (Y/N)?
```

Program notes

(1) Lines 10–30 set up title
(2) Lines 40–60 enter experimental observations
(3) Line 70 calculates base e reaeration coefficient
(4) Lines 80–90 calculate and print base 10 coefficient
(5) Lines 100–120 offer predictions for other conditions
(6) Line 140 enters new initial dissolved oxygen
(7) Line 160 prints table heading
(8) Line 170 sets up loop
(9) Line 180 calculates dissolved oxygen deficit
(10) Line 190 calculates dissolved oxygen concentration
(11) Line 200 prints results
(12) Line 210 returns loop
(13) Lines 220–240 offer another initial DO if desired
(14) Lines 250–270 offer another experiment if desired

Example 4.3 DOSAG: dissolved oxygen sag program

The Streeter Phelps equation (Equation 4.7) can be used to predict dissolved oxygen concentrations in a stream. Write a program which will be able to provide a stepwise solution to a river system with a series of tributaries and effluent discharges. Use it to solve the following problem.

A wastewater effluent of 1 m^3/s, DO 4 mg/l and BOD 20 mg/l discharges to a stream with a flow of 3 m^3/s, saturated with DO and with a BOD of 2 mg/l. At 1 day time of travel below the effluent discharge there is a confluence with a tributary having a flow of 1.5 m^3/s, DO 8 mg/l and BOD 2 mg/l. A further 1.5 days time of travel below the confluence there is another wastewater effluent discharge of 4 m^3/s, DO 6 mg/l and BOD 20 mg/l. Determine the DO deficit at a point 2 days time of travel below the second effluent discharge. The BOD rate constant for all reaches of the river is 0.1/day and the reaeration coefficients are: from first discharge to confluence 0.40/day, from confluence to second discharge 0.35/day and from second discharge onwards 0.30/day. The DO saturation concentration is 9.1 mg/l.

```
10 CLS
20 REM DOSAG
30 PRINT"Dissolved Oxygen Sag Program"
40 INPUT"Enter initial flow, DO (mg/l), BOD (mg/l)";Q1,D1,B1
50 INPUT"Enter next flow, DO (mg/l), BOD (mg/l)";Q2,D2,B2
60 BI=(Q1*B1+Q2*B2)/(Q1+Q2)
70 DI=(Q1*D1+Q2*D2)/(Q1+Q2)
80 INPUT"Enter BOD rate constant (base 10 per day)";K1
90 INPUT"Enter reaeration constant (base 10 per day)";K2
100 INPUT"Enter saturation DO (mg/l)";DS
110 DA=DS-DI:KB=K1*2.3026:KA=K2*2.3026
120 LA=BI/(1-EXP(-KB*5))
130 INPUT"Enter time of flow in section (d)";TF
140 INPUT"Enter number of segments required in section";N
150 CLS
160 PRINT"Time","Deficit"," DO "," BOD "
170 PRINT" (d)"," (mg/l)","(mg/l)","(mg/l)"
180 FOR T=0 TO TF STEP TF/N
190 DT=(KB*LA/(KA-KB))*(EXP(-KB*T)-EXP(-KA*T))+DA*EXP(-KA*T)
200 D=DS-DT
210 LT=LA*EXP(-KB*T)
220 BT=LT*(1-EXP(-KB*5))
230 PRINT T,DT,D,BT
240 IF DT>DS THEN PRINT"ZERO DO...BOD TOO HIGH":END
250 NEXT T
260 ON ERROR GOTO 410
270 TC=LOG((1-(DA*(KA-KB)/(LA*KB)))*KA/KB)*(1/(KA-KB))
280 IF TC<=0 THEN 410
290 IF TC>TF THEN 420
300 DC=(KB/KA)*LA*EXP(-KB*TC)
310 PRINT"Minimum DO=";DS-DC;"mg/l at";TC;"days"
320 PRINT
330 INPUT"Another section (Y/N)";Q$
340 IF Q$="Y" OR Q$="y" THEN 350 ELSE END
350 Q1=Q1+Q2
360 D1=D
370 B1=BT
380 CLS
390 GOTO 50
400 END
410 TC=0:DC=DA:GOTO 310
420 TC=TF:DC=DT:GOTO 310
430 END
```

```
Dissolved Oxygen Sag Program
Enter initial flow, DO (mg/l), BOD (mg/l)? 3,9.1,2
Enter next flow, DO (mg/l), BOD (mg/l)? 1,4,20
Enter BOD rate constant (base 10 per day)? .1
Enter reaeration constant (base 10 per day)? .4
Enter saturation DO (mg/l)? 9.1
Enter time of flow in section (d)? 1
Enter number of segments required in section? 4

Time            Deficit          DO             BOD
 (d)             (mg/l)         (mg/l)          (mg/l)
 0               1.275           7.825001        6.5
 .25             1.487225        7.612775        6.136394
 .5              1.629261        7.47074         5.793127
```

```
 .75            1.717027       7.382974       5.469063
 1              1.763088       7.336913       5.163126
Minimum DO= 7.336913 mg/l at 1 days

Another section (Y/N)? Y

Enter next flow, DO (mg/l), BOD (mg/l)? 1.5,8,2
Enter BOD rate constant (base 10 per day)? .1
Enter reaeration constant (base 10 per day)? .35
Enter saturation DO (mg/l)? 9.1
Enter time of flow in section (d)? 1.5
Enter number of segments required in section? 5

Time            Deficit          DO             BOD
 (d)             (mg/l)         (mg/l)         (mg/l)
 0              1.582245       7.517755        4.300455
 .3             1.614808       7.485193        4.013416
 .6             1.615523       7.484477        3.745536
 .9000001       1.592889       7.507112        3.495536
 1.2            1.553469       7.546532        3.262223
 1.5            1.502312       7.597689        3.044482
Minimum DO= 7.481434 mg/l at .4540246 days

Another section (Y/N)? Y

Enter next flow, DO (mg/l), BOD (mg/l)? 4,6,20
Enter BOD rate constant (base 10 per day)? .1
Enter reaeration constant (base 10 per day)? .3
Enter saturation DO (mg/l)? 9.1
Enter time of flow in section (d)? 2
Enter number of segments required in section? 4

Time            Deficit          DO             BOD
 (d)             (mg/l)         (mg/l)         (mg/l)
 0              2.175022       6.924979        10.18365
 .5             2.904809       6.195192        9.076178
 1              3.273012       5.826988        8.089147
 1.5            3.401378       5.698623        7.209454
 2              3.37434        5.725661        6.425428
Minimum DO= 5.693435 mg/l at 1.635513 days

Another section (Y/N)? N
```

Program notes

(1) Lines 10–30 set up title
(2) Lines 40–50 enter flow, DO and BOD data for both streams
(3) Lines 60–70 mass balance calculations for BOD and DO
(4) Lines 80–100 enter rate constants and saturation DO
(5) Line 110 calculates initial DO deficit and base e constants
(6) Line 120 calculates initial ultimate BOD
(7) Lines 130–140 enter section details
(8) Lines 150–170 print table headings
(9) Line 180 sets up loop for specified segments
(10) Line 190 DO deficit by Streeter Phelps equation
(11) Lines 200–230 calculate and print DO and BOD values
(12) Line 240 ends program if excessive pollution gives negative DO levels

(13) Line 250 returns loop
(14) Line 260 error trap for negative value of critical time
(15) Line 270 calculates critical time
(16) Lines 280–290 allow for critical time outside section
(17) Line 300 calculates critical DO
(18) Line 310 prints critical time and DO
(19) Lines 320–340 offer another section if desired
(20) Lines 350–370 new initial values for next section
(21) Line 380 returns program to next input stream
(22) Lines 410–420 set correct values after error trap (if critical time is < 0 then TC = 0; or if critical time is $>$ TF then TC = TF)

PROBLEMS

(4.1) A water abstraction is made from a river some days time of flow below an effluent discharge which contains both conservative and non-conservative pollutants. Limits have been set at the abstraction point for the acceptable levels of these pollutants in the abstracted water. Write a program which will tabulate the allowable levels of both types of pollutant in the effluent to satisfy the downstream constraints for a range of river flows. Allowance must be made for the presence of background concentrations of the pollutants in the river before the effluent discharge point and for varying the decay constant in the case of non-conservative pollutants.

(4.2) An empirical relationship for prediction of the reaeration coefficient in flowing water is

$$k_2 = cv^n/h^m \tag{4.12}$$

where v = mean velocity (m/s)
h = mean depth (m)
c, n and m are constants (typical values: 2.1, 0.9, 1.7 respectively)

Write a program which will calculate k_2 values and print the values as a matrix for five velocities and five depths of flow.

(4.3) Most rate coefficients are influenced by temperature, and the relationship is commonly expressed as

$$k_T = k_{20}\,\theta^{(T-20)} \tag{4.13}$$

where k_T = value of coefficient at T°C
k_{20} = value of coefficient at 20°C
θ = a constant, the value of which depends upon the reaction

For biochemical reactions, such as BOD, θ is often taken as 1.047, whereas for physical reactions, like gas transfer, a lower value of around 1.02 is common.

Write a program which tabulates k values for a temperature range of 1 to 30°C with a user-selected value for θ.

(4.4) Modify the program in Worked Example 4.3 to enable solutions to be obtained taking into account the effects of temperature variations on the saturation dissolved oxygen levels and the values of the rate constants k_1, k_2.

Temp. (°C)	0	5	10	15	20	25	30
Satn DO (mg/l)	14.6	12.8	11.3	10.2	9.1	8.4	7.6

FURTHER READING

Tebbutt, T. H. Y., *Principles of Water Quality Control*, Chapter 7, 3rd edn, Pergamon Press, 1983.

Velz, C. J., *Applied Stream Sanitation*, Wiley-Interscience, 1970.

Chapter 5

Physical treatment processes

Most sources of water will require some treatment to remove undesirable constituents or simply as a precautionary measure to deal with accidental contamination. Similarly, most wastewaters need considerable purification before they can be safely discharged into the surrounding environment. With waters and wastewaters there is likely to be more than one undesirable constituent, so it is often necessary to utilize different treatment processes in series to achieve the desired product quality. This means that most treatment plants are a system of unit processes or operations linked together in the most effective way to produce the desired quality changes. In order to select the most appropriate combination of processes, a designer must be aware of the characteristics of the impurities which are to be removed and the principles of the available treatment processes.

Although the characteristics of waters and wastewaters vary widely, it is possible to classify the main types of impurities found into five groups.

1. Floating or large suspended solids – leaves, twigs, fish, paper, rags, grit.
2. Fine suspended and colloidal solids – clay, silt, micro-organisms, proteins.
3. Dissolved solids – alkalinity, hardness, salts, organics, metals.
4. Dissolved gases – carbon dioxide, hydrogen sulphide.
5. Immiscible substances – oils and greases.

In certain circumstances it is necessary to add materials as part of the treatment process; this could involve the use of coagulants in water treatment, the addition of chlorine for disinfection of water, or the aeration of sewage in biological oxidation. The divisions betwen different types of impurities are not, of course, sharply defined, and there may often be a degree of overlap in their properties. Nevertheless, the recognition of different types of impurities is useful as a means of characterizing the material which has to be handled and in giving some indication of the method of treatment which would be most appropriate. There are three main classes of treatment methods, each having several different forms:

1. Physical processes which depend upon purely physical properties of the impurity such as size and density. Typical processes are straining, sedimentation and filtration.
2. Chemical processes which depend upon the chemical properties of the impurity or an added reagent such as solubility and degree of ionization. Typical processes are coagulation, precipitation and ion exchange.
3. Biological processes which use biochemical reactions to remove impurities which are utilized as sources of food for microorganisms housed in a suitable reactor. These processes, which may be aerobic or anaerobic, are mainly used for dealing with organic impurities in wastewater treatment. Typical processes are activated sludge, biological filtration, oxidation ponds and anaerobic digestion.

The remainder of this chapter is concerned with physical treatment processes, with the other forms of treatment being covered in succeeding chapters.

ESSENTIAL THEORY

5.1 Sedimentation

Solids suspended in a fluid will settle at a velocity dependent upon their size and density and the density and viscosity of the fluid. The process occurs naturally in the deposition of suspended matter when a river enters a lake, because as the horizontal velocity is reduced, there is more opportunity for the solids to reach the bottom. Sedimentary rocks are another natural example of settlement processes, although here the time-scale for deposition is very long, so that small particles are deposited. Many waters and wastewaters contain suspended solids which can be removed by sedimentation, although it must be appreciated that as the settling velocity of particles becomes smaller with reducing size and density, the process reaches its practical limit because long retention times are uneconomic.

Suspensions may be composed of discrete particles, which have a fixed rigid surface and do not readily agglomerate, or of flocculent particles, which have non-rigid surfaces and agglomerate when brought into contact with each other. As demonstrated in Figure 5.1, a suspension of discrete particles has a constant settling velocity with depth, whereas a flocculent suspension shows an increasing settling velocity with depth due to the growth in size of individual particles by collisions with less rapidly settling particles.

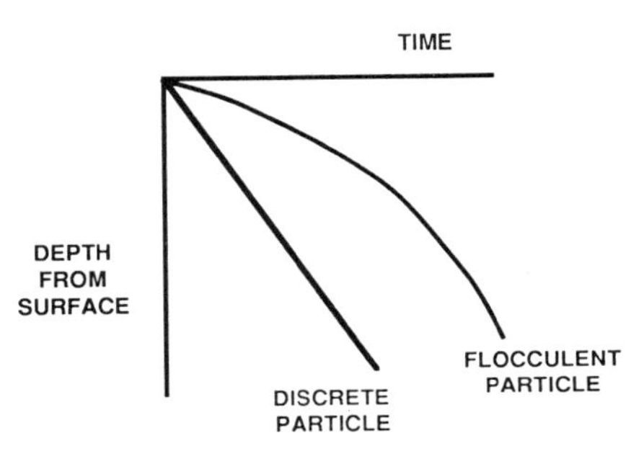

Figure 5.1 Settling characteristics of discrete and flocculent particles

Simple settling theory is based on the behaviour of discrete particles, although in many waters and wastewaters the bulk of solids may well be flocculent in nature so that in practice the theory requires some modification. When a discrete particle is placed in a fluid of lower density it will accelerate until a terminal velocity is reached when the gravitational force is equal and opposite to the frictional drag force:

$$\text{gravitational force} = (\rho_s - \rho_w)gV \tag{5.1}$$

where ρ_s = density of particle
ρ_w = density of fluid
g = acceleration due to gravity
V = volume of particle

Frictional force is a function of settling velocity, size of particle, and density and viscosity of fluid, and by dimensional analysis it can be shown that

$$\text{frictional force} = C_D A_c \rho_w v_s^2/2 \tag{5.2}$$

where C_D = Newton's drag coefficient
A_c = cross-sectional area of particle
v_s = terminal settling velocity

For spherical particles with Reynolds Number (R_N) ≤ 1

$$C_D = 24/R_N \tag{5.3}$$

For spherical particles with $R_N > 1$

$$C_D = 24/R_N + 3/R_N^{0.5} + 0.34 \tag{5.4}$$

By equating gravitational and frictional forces an expression for the terminal settling velocity can be obtained.

$$v_s = [2gV(\rho_s - \rho_w)/C_D A_c \rho_w]^{0.5} \tag{5.5}$$

For spheres $A_c = \pi d^2/4$, $V = \pi d^3/6$

$$v_s = [4gd(\rho_s - \rho_w)/3C_D\,\rho_w]^{0.5} \tag{5.6}$$

In laminar flow conditions ($R_N \leqslant 1$)

$$C_D = 24\nu/v_s d \tag{5.7}$$

where ν = kinematic viscosity

Hence Stoke's Law is

$$v_s = (gd^2/18\nu)[(\rho_s - \rho_w)/\rho_w] \tag{5.8}$$

A basic understanding of the process of sedimentation can be obtained from a consideration of the concept of an ideal settling basin as illustrated in Figure 5.2. The assumptions made for such an idealized situation include quiescent conditions in the settling zone, uniform flow across the settling zone, uniform solids concentration both vertically and horizontally at the entrance to the settling zone, and no resuspension from the sludge zone. Given these idealized conditions, a discrete particle which enters at the top of the settling zone and just reaches the bottom at the outlet end of the settling zone will have a settling velocity v_0.

$$v_0 = \text{distance settled/time} = h_0/t_0 \tag{5.9}$$

But

$$t_0 = \text{tank volume/flow rate} = Ah_0/Q \tag{5.10}$$

Thus

$$v_0 = Q/A \tag{5.11}$$

The expression Q/A is termed the surface overflow rate, and it can be seen that for discrete particles, removal by sedimentation is controlled by the surface area of the tank and is independent of depth. For particles with a settling velocity v_s less than v_0, some removal will occur depending on the depth at which they enter the tank. Removal will occur if particles enter the tank at a distance from the bottom of the settling zone of h or less where $h = v_s t_0$. Thus the proportional removal of particles with $v_s < v_0$ will be in the ratio of v_s/v_0.

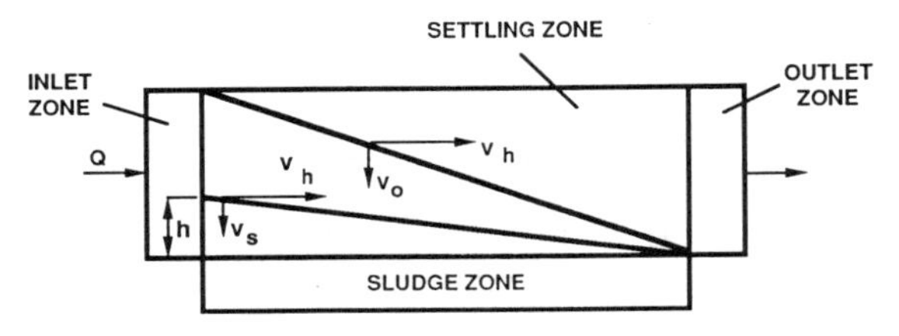

Figure 5.2 The ideal settling basin

Experimental measurement of the settling characteristics of a suspension of discrete particles is carried out using a vertical column, about 3 m high, provided with a tapping point at a known depth from the surface. The column is filled with the suspension and thoroughly mixed and then allowed to settle under quiescent conditions. Samples are taken from the tapping point at depth d below the surface at the start of the test and at known time intervals. None of the suspended solids in the samples collected will have velocities which would have carried them past the sampling point in the time since the start of the test, i.e.

$$v_s \text{ for each sample} \leqslant d/t_1,\ d/t_2,\ d/t_3, \text{ etc.}$$

These results can then be plotted in the form shown in Figure 5.3 to give a settling characteristic curve for the suspension which is analogous to a sieve analysis plot based on particle size. The settling characteristic curve can be used to predict the performance of an ideal settling tank when fed with a suspension of discrete particles. For a horizontal flow unit, all particles with $v_s \geqslant v_0$ will be removed with an additional removal of particles having $v_s < v_0$ in the ratio v_s/v_0. Thus the overall percentage removal is given by

$$E = 100 - p_0 + \int_0^{p_0} (v_s/v_0)\,\mathrm{d}p \qquad (5.12)$$

For a vertical flow tank only those particles with $v_s \geqslant v_0$ will be removed, since others will be washed out of the tank. Thus in this case

$$E = 100 - p_0 \qquad (5.13)$$

With flocculent suspensions, prediction of performance is more difficult, since the depth of the settling tank affects the agglomeration of the suspension and hence its settling velocity. In addition it is important to realize that the ideal settling basin is subject to some modification in real life, due to such factors

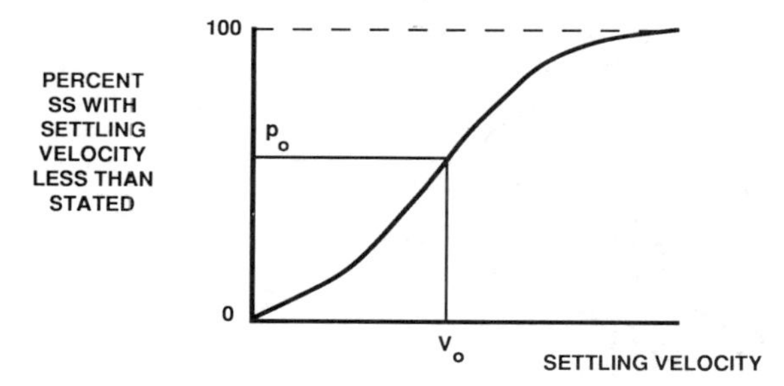

Figure 5.3 Typical settling characteristic curve

as hydraulic turbulence, density currents and wind action. The retention time of a settling basin is therefore almost always significantly less than the theoretical value.

5.2 Flocculation

With small suspended particles the calculated settling velocities become so small that sedimentation is clearly impracticable, and for most purposes sedimentation is not an appropriate treatment process for particles smaller than about 50 μm. If particles are flocculent it is possible to utilize this property and encourage agglomeration of fine material so that settleable solids are produced. Such agglomeration is brought about largely by the creation of velocity gradients in the suspension which cause collisions between particles. The number of collisions in a suspension is proportional to the velocity gradient, so that within limits the degree of agglomeration increases with the level of mixing to which the suspension is exposed. As flocculation produces larger and larger particles, the suspension then becomes more sensitive to the effects of shear so that high velocity gradients will eventually result in break-up of large floc particles.

Velocity gradients can be created either by hydraulic turbulence caused by baffles or by mechanically driven paddles. The latter are more flexible in use since the degree of flocculation can be easily adjusted. Figure 5.4 illustrates such a system.

By considering the behaviour of an element of fluid exposed to a velocity gradient it can be shown that the relationship between velocity gradient G and power input per unit volume P is

$$P = \mu G^2 \tag{5.14}$$

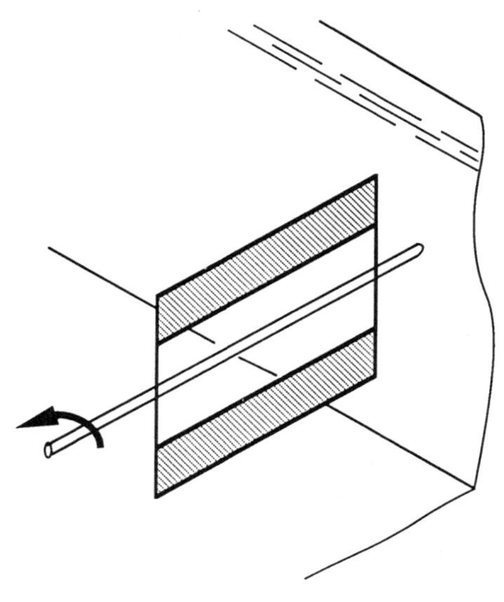

Figure 5.4 Paddle-type flocculation system

For a paddle flocculator

power = drag on paddles × velocity of paddles relative to fluid

$$P = C_D A v^3/2V \qquad (5.15)$$

where v = velocity of paddles relative to fluid

V = volume of tank

For most flocculent suspensions in water and wastewater treatment, an appropriate value of G is usually 20–75 s^{-1}.

5.3 Flow through porous media

Filtration through a bed of granular media, usually sand, is a common method of potable water treatment and is also used for 'polishing' of wastewater effluents in situations where a high-quality discharge is required. Although superficially simple, filtration is actually a highly complex process involving a number of transport and attachment mechanisms. As a result, a deep filter bed can remove particles much smaller than the voids between the individual bed grains. Theoretical consideration of the behaviour of a filter bed as it becomes clogged is thus highly complex, and this section will only consider the basic hydraulic factors which influence the behaviour of filters and granular beds used in activated carbon treatment to remove trace organics and in ion-exchange processes. Sand filters (Figure 5.5) used in potable water treatment, which are by far the most common application of porous media beds, are of two types: slow filters with a low hydraulic loading and little penetration of solids, and rapid filters with much higher hydraulic loadings and deep penetration of solids. Slow filters are cleaned at infrequent intervals by removing the clogged surface layer, whereas rapid filters need frequent cleaning which is achieved by a backwashing operation in which filtered water is pumped up through the bed to expand it.

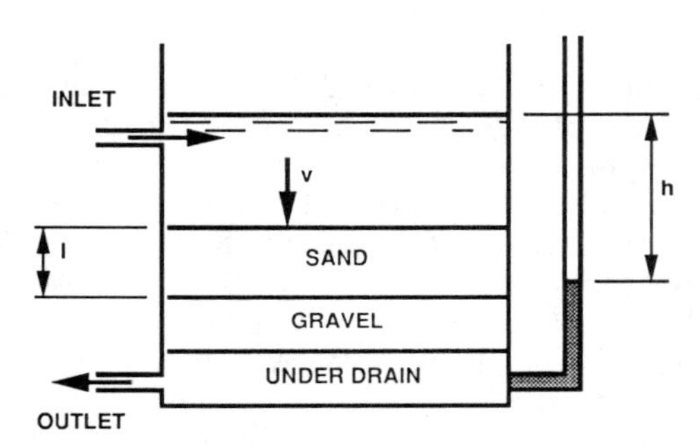

Figure 5.5 Schematic of rapid gravity filter

The hydraulics of filtration may be considered as analogous to flow through small pipes and to the resistance offered by a fluid to settling particles. An empirical approach to the situation was developed by Rose from experimental work which showed that the head loss per unit bed depth was proportional to v^2/gd and to $1/f^4$. By introducing Newton's drag coefficient, which appears in Section 5.1, it was possible to develop the following expression:

$$\frac{h}{l} = 1.067 C_d v^2/(gd\psi f^4) \tag{5.16}$$

where d = bed grain diameter
f = porosity of bed (volume of voids/total volume)
ψ = particle shape factor (relative surface areas of a sphere of the same volume as bed grain)

(for spheres $\psi = 1$ and $d = 6V/A$)

The Rose equation applies to a bed composed of single-size

grains but in many cases beds are made up from a graded material and the overall head loss must be determined by an arithmetic integration for each size range. In the case of a slow sand filter, the grading is uniform throughout the depth of the bed, but in the case of rapid filters, the backwashing produces a stratified bed with the smallest particles at the top and the coarsest at the bottom.

Filter cleaning is necessary when the output flow rate or required filtrate quality can no longer be maintained. With a slow filter the solids penetration is only superficial, so that cleaning can be achieved by scraping off the top few centimetres of the bed at intervals of several months. The rapid filter becomes quickly clogged throughout most of its depth and requires cleaning at intervals of every one or two days in most installations. This is carried out by introducing water and compressed air into the base of the bed at a rate sufficient to produce a fluidized state. This produces a scouring action and the expansion of the voids allows the deposits to be removed in the washwater flow. As backwash water is admitted to the bottom of the bed, the head loss increases, and as the velocity is increased, the bed will eventually begin to expand. As the expansion continues further, the head loss within the bed becomes essentially constant and is equal to the weight of the media in water. With reference to Figure 5.6, it can be shown that when a fluidized bed has been produced, the upward force produced by the backwash water is equal and opposite to the weight of the expanded bed, i.e.

$$hg\rho_w = l_e(\rho_s - \rho_w)g(1 - f_e) \tag{5.17}$$

or

$$h/l_e = [(\rho_s - \rho_w)/\rho_w](1 - f_e) \tag{5.18}$$

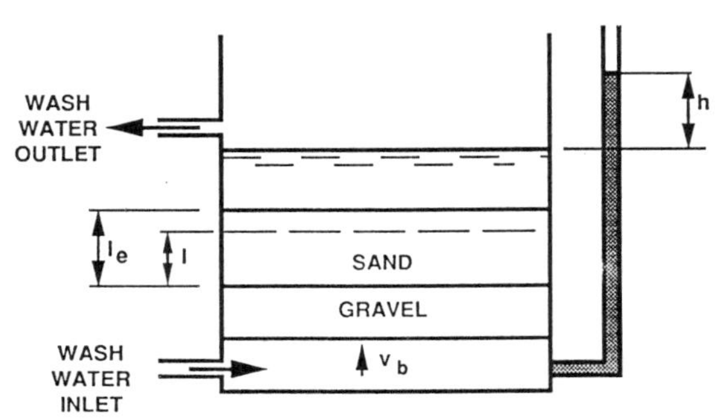

Figure 5.6 Gravity filter under backwash conditions

This expression gives the head loss per unit depth for an expanded bed with single-size grains. The grains are kept in suspension because of the drag exerted on them by the upward flow of water, and it has been found by experiment that the velocity (v_b) necessary to expand a bed of granular material to a specific porosity is related to the settling velocity of individual grains:

$$f_e = (v_b/v_s)^{0.22} \qquad (5.19)$$

Thus by continuity, since the unexpanded and expanded beds contain the same material,

$$(1 - f)l = (1 - f_e)l_e \qquad (5.20)$$

i.e.

$$l_e/l = (1 - f)/(1 - f_e)$$

and

$$l_e/l = (1 - f)/[1 - (v_b/v_s)^{0.22}] \qquad (5.21)$$

For graded beds an arithmetic integration is necessary, since each grain size is carried into suspension in succession as the drag force becomes equal to the settling force for that particular grain size.

WORKED EXAMPLES

Example 5.1 DISPAR: settling velocity of discrete particles

Develop a program which will calculate the settling velocity of a specified particle at a given temperature. The program should check that the expression used to calculate the settling velocity is appropriate for the Reynolds Number which the settling particle will have.

Determine the settling velocity of a spherical discrete particle of diameter 0.06 mm and mass density 2650 kg/m^3. The kinematic viscosity (ν) of water varies with temperature as shown below.

Temp. (°C)	*Viscosity* (10^{-6}m^2/s)
5	1.519
10	1.306
15	1.139
20	1.003
25	0.893
30	0.800

```
10 CLS
20 REM DISPAR
30 PRINT"Settling Velocity of Discrete Particles in Water"
40 INPUT"Enter particle diameter (mm)";D
50 INPUT"Enter mass density of particle (kg/cu.m)";MD
60 INPUT"Select temp: 5, 10, 15, 20, 25 or 30 deg C";T
70 IF T=5 THEN NU=1.519E-06
80 IF T=10 THEN NU=1.306E-06
90 IF T=15 THEN NU=1.139E-06
100 IF T=20 THEN NU=1.003E-06
110 IF T=25 THEN NU=8.93E-07
120 IF T=30 THEN NU=.0000008
130 VS=(9.810001/18)*(D*.001)^2* ((MD-1000)/1000)/NU
140 PRINT
150 PRINT"Settling velocity at";T;"deg C =";VS;"m/s"
160 RN=VS*D*.001/NU
170 PRINT"Reynolds Number=";RN
180 Q$=""
190 IF RN<=1 THEN INPUT"Another calculation (Y/N)";Q$
200 IF Q$="Y" OR Q$="y" THEN 40
210 IF Q$="N" OR Q$="n" THEN END
220 PRINT"RN too high for Stokes's Law"
230 PRINT
240 PRINT"Recalculating using Newton's Law"
250 PRINT
260 CD=24/RN+3/SQR(RN)+.34
270 VM=SQR(4*9.810001*(MD-1000)*D*.001/(3*CD*1000))
280 RM=VM*D*.001/NU
290 IF RM/RN>.95 THEN 310
300 RN=RM: GOTO 270
310 PRINT"Settling velocity at ";T;"deg C =";VM;"m/s"
320 PRINT"Reynolds Number=";RM
330 PRINT
340 INPUT"Another calculation (Y/N)";Q$
350 IF Q$="Y" OR Q$="y" THEN 40
360 END
```

```
Settling Velocity of Discrete Particles in Water
Enter particle diameter (mm)? .06
Enter mass density of particle (kg/cu.m)? 2650
Select temp: 5, 10, 15, 20, 25 or 30 deg C? 20

Settling velocity at 20 deg C = 3.227618E-03 m/s
Reynolds Number= .1930778
Another calculation (Y/N)?
```

Program notes

(1) Lines 10–30 set up title
(2) Lines 40–60 enter particle details and temperature
(3) Lines 70–120 kinematic viscosity data
(4) Lines 130–150 calculate and print Stokes Law velocity
(5) Lines 160–170 calculate and print Reynolds Number
(6) Line 180 initializes Q$ for repeat calculations
(7) Lines 190–210 control program if Reynolds Number is OK
(8) Lines 220–250 indicate that initial Reynolds Number is incorrect

(9) Line 260 uses initial Reynolds Number to obtain an approximate value of C_D from Equation 5.4
(10) Line 270 calculates settling velocity from Equation 5.6
(11) Line 280 calculates modified Reynolds Number
(12) Line 290 if modified Reynolds Number is close to previous value, program proceeds to print
(13) Line 300 if modified Reynolds number is not close to previous value, new Reynolds Number is used in a repeated calculation. This will continue until 290 is satisfied
(14) Lines 310–320 print out results
(15) Lines 330–350 offer another calculation if desired

Example 5.2 IDEALSED: performance of an ideal settling tank

Write a program to calculate the surface overflow rate of a settling tank given its dimensions and the throughput. The program should cater for both circular and rectangular tanks. Provision should be made for calculation of the removal performance for particles with a settling velocity less than the surface overflow rate.

Determine the surface overflow rate of a rectangular tank 25 m long by 6 m wide and 2 m deep when operating with a flow of 0.05 m^3/s. Hence calculate the theoretical removal in this tank of discrete particles with a settling velocity of 0.0002 m/s.

```
10 CLS
20 REM IDEALSED
30 PRINT"Performance of an Ideal Settling Tank"
40 INPUT"Enter flow to be treated (cu.m/s)";Q
50 INPUT"Is tank circular or rectangular (C/R)";Q$
60 IF Q$="C" OR Q$="c" THEN 100
70 INPUT"Enter length, width, depth (m)";L,W,D
80 SA=L*W
90 GOTO 120
100 INPUT"Enter diameter, depth (m)";DI,D
110 SA=DI^2*3.14/4
120 SL=Q*60*60*24/SA
130 T=D/SL*24
140 PRINT
150 PRINT"Surface overflow rate=";SL;"m/d"
160 PRINT"retention time=";T;"h"
170 VO=SL*1000/86400!
180 PRINT"Tank will remove all particles with Vs>";VO;"mm/s"
190 PRINT
200 PRINT"Removal of particles with velocities lower"
210 PRINT"than Vs can now be calculated"
220 INPUT"Enter particle velocity (mm/s)";VS
230 R=100*VS/VO
240 IF R>100 THEN R=100
250 PRINT
260 PRINT"Removal of particles with Vs=";VS;"=";R;"percent"
270 PRINT
280 INPUT"Another calculation for same tank (Y/N)";Q$
```

```
290 IF Q$="Y" OR Q$="y" THEN 220
300 PRINT
310 INPUT"Calculation for a new tank (Y/N)";Q$
320 IF Q$="Y" OR Q$="y" THEN 10
330 END
```

```
Performance of an Ideal Settling Tank
Enter flow to be treated (cu.m/s)? .05
Is tank circular or rectangular (C/R)? R
Enter length, width, depth (m)? 25,6,2

Surface overflow rate= 28.8 m/d
retention time= 1.666667 h
Tank will remove all particles with Vs> .3333334 mm/s

Removal of particles with velocities lower
than Vs can now be calculated
Enter particle velocity (mm/s)? .2

Removal of particles with Vs= .2 = 60 percent

Another calculation for same tank (Y/N)? N

Calculation for a new tank (Y/N)?
```

Program notes

(1) Lines 10–30 set up title
(2) Lines 50–60 query for shape of tank
(3) Lines 70–80 rectangular tank input and calculation
(4) Line 90 directs program to skip circular input
(5) Lines 100–110 circular tank input and calculation
(6) Line 120 calculates surface overflow rate
(7) Line 130 calculates retention time
(8) Lines 140–160 print results
(9) Lines 170–180 convert overflow rate to particle settling velocity and print
(10) Lines 190–210 set up subtitle
(11) Line 220 enters particle velocity
(12) Lines 230–260 calculate removal and print result
(13) Lines 270–290 offer another velocity if desired
(14) Lines 300–320 offer another tank if desired

Example 5.3 FLOC: flocculation calculations

Flocculation is frequently provided by means of motor-driven paddles, and a program is required to calculate the power requirements and velocity gradient for specified configurations and conditions.

A flocculation tank is 10 m long, 3 m wide and 3 m deep with a design flow of 0.046 m^3/s. There are three paddle wheels each with two blades 2.4 m by 0.35 m, the centre line of the blades being 1 m from the shaft which is at mid-depth of the tank. The speed of rotation of the paddles is 3.2 rev/min and C_D for the blades is 1.8. The velocity of the blades relative to the water is 0.75 times the rotational velocity. ν is $1.011 \times 10^{-6}\, m^2/s$.

```
10 CLS
20 REM FLOC
30 PRINT"Flocculation Calculations"
40 INPUT"Enter tank dimensions L, W, D (m)";L,W,D
50 INPUT"Enter flow rate (cu.m/s)";Q
60 DT=L*W*D/(Q*60)
70 IF DT<15 THEN PRINT"Detention time < 15 min":GOTO 40
80 INPUT"Enter number of paddles, blades per paddle";P,N
90 INPUT"Enter blade L, W, and radius (m)";BL,BW,BR
100 IF P*2*BR-1>L THEN PRINT"Paddles will not fit!":GOTO 80
110 IF BL>W THEN PRINT"Paddles wider than tank!":GOTO 90
120 IF BR>.45*D THEN PRINT"Paddles too deep!":GOTO 90
130 INPUT"Enter revolutions/min of paddles";RP
140 CD=1.8:NU=1.011E-06
150 PV=2*3.142*BR*RP/60
160 DV=PV*.75
170 PA=P*N*BL*BW
180 PR=CD*PA*DV^3*1000/2
190 G=SQR(CD*PA*DV^3/(2*NU*L*W*D))
200 PRINT
210 PRINT"Power requirement=";PR;"watts"
220 PRINT"Velocity gradient=";G;"m/s m"
230 PRINT"Detention time=";DT;"min"
240 PRINT
250 INPUT"Another set of calculations (Y/N)";Q$
260 IF Q$="Y" OR Q$="y" THEN 10
270 END
```

```
Flocculation Calculations
Enter tank dimensions L, W, D (m)? 10,3,3
Enter flow rate (cu.m/s)? 0.046
Enter number of paddles, blades per paddle? 3,2
Enter blade L, W, and radius (m)? 2.4,0.35,1
Enter revolutions/min of paddles? 3.2

Power requirement= 72.03795 watts
Velocity gradient= 28.13739 m/s m
Detention time= 32.6087 min

Another set of calculations (Y/N)?
```

Program notes

(1) Lines 10–30 set up title
(2) Lines 40–50 enter tank details
(3) Line 60 calculates detention time
(4) Line 70 if detention time too short, returns to 40 for revised input
(5) Lines 80–90 enter paddle details

(6) Lines 100–120 check for paddles that do not fit tank
(7) Line 130 enters paddle speed
(8) Line 140 values for C_D and kinematic viscosity
(9) Line 150 calculates paddle velocity
(10) Line 160 calculates relative velocity of paddle
(11) Line 170 calculates paddle area
(12) Line 180 calculates power requirement from Equation 5.15
(13) Line 190 calculates velocity gradient from Equation 5.14
(14) Lines 200–230 print results
(15) Lines 240–260 offer another calculation if desired

Example 5.4 FILTER: deep-bed filter head losses using Rose's equation

Write a program to determine the head losses in a bed of granular medium using the Rose formula for a range of filtration rates.

A filter bed 900 mm deep is made up of 0.5 mm worn sand and has a porosity of 40%. Determine the head loss at filtration rates of 50–150 m/day.

```
10 CLS
20 REM FILTER
30 PRINT"Deep-Bed Filter Headlosses Using Rose's Equation"
40 PRINT
50 INPUT"Depth (mm), grain size (mm), porosity (%)";B,D,F
60 IF F>99 THEN PRINT"Porosity cannot > 100%!":GOTO 40
70 INPUT"Enter shape factor: new sand 0.7, worn sand 0.9";S
80 INPUT"Enter min, and max filtration rates (m/d)";F1,F2
90 PRINT"Program uses kinematic viscosity of water at 20 C"
100 PRINT"For another viscosity enter A, otherwise ENTER"
110 INPUT;A$
120 PRINT
130 IF A$="A" OR A$="a" THEN PRINT"Edit l 140 and run":END
140 NU=1.003E-06
150 PRINT
160 PRINT"Filtrn Rate","Head Loss","Head/Depth"
170 PRINT" (m/d)      "," (m)    ","     (%)     "
180 FOR FR=F1 TO F2 STEP INT((F2-F1)/10)
190 RN=FR/(60*60*24)*D*.001/NU
200 CD=24/RN+3/SQR(RN)+.34
210 H=B*.001*1.067*CD*(FR/86400!)^2/(S*9.8*D*.001*(F*.01)^4)
220 R=H/B*100000!
230 PRINT FR,H,R
240 NEXT FR
250 PRINT
260 INPUT"Another set of calculations (Y/N)";Q$
270 IF Q$="Y" OR Q$="y" THEN 10
280 END
```

```
Deep-Bed Filter Headlosses Using Rose's Equation

Depth (mm), grain size (mm), porosity (%)? 900,.5,40
```

```
Enter shape factor: new sand 0.7, worn sand 0.9? .9
Enter min, and max filtration rates (m/d)? 50,150
Program uses kinematic viscosity of water at 20 C
For another viscosity enter A, otherwise ENTER
?

Filtrn Rate    Head Loss      Head/Depth
 (m/d)           (m)             (%)
 50             .2538679       28.20755
 60             .3066963       34.07737
 70             .3600387       40.0043
 80             .4138625       45.98472
 90             .4681416       52.01573
 100            .5228542       58.09491
 110            .5779818       64.2202
 120            .6335092       70.38991
 130            .6894223       76.60248
 140            .7457087       82.85652
 150            .8023582       89.15091

Another set of calculations (Y/N)? N
```

Program notes

(1) Lines 10–40 set up title
(2) Line 50 enters bed particulars
(3) Line 60 traps error in entry of porosity
(4) Lines 70–80 enter appropriate shape factor
(5) Lines 90–140 option to change kinematic viscosity
(6) Lines 150–170 set up headings for table
(7) Line 180 sets up loop
(8) Line 190 calculates Reynolds Number
(9) Line 200 calculates C_D
(10) Line 210 calculates head loss from Equation 5.16
(11) Line 220 calculates head loss as percentage of depth
(12) Line 230 prints results
(13) Line 240 returns loop
(14) Lines 250–270 offer another calculation if desired

PROBLEMS

(5.1) Write a program to take the results from a laboratory settling column test and process them in a form suitable for further analysis. The basic results will be in the form of suspended solids filter paper weights for samples collected at known time intervals from a sampling point in the column. Because of the removal of samples from the settling column, the depth of the sampling point from the liquid surface will decrease with time and the program should take into account this factor. The output of the program should be a table of velocities and percentages of suspended solids with velocities less than or equal to the particular velocities.

(5.2) Use the program written for the problem above and Worked Example 5.2 as a basis for a program to determine the removal of suspended solids in an ideal horizontal-flow sedimentation tank. Equation 5.12 defines this removal performance.

(5.3) Write a program which produces the main dimensions and paddle configuration for a mechanical flocculation basin. The user will input rate of flow, velocity gradient and depth of tank. The program should interact with the user to offer the choice of length : width ratio, and number and type of paddles. It should prevent the adoption of unsatisfactory dimensions and ensure that the product of velocity gradient and retention time is within acceptable limits. The program output should list tank dimensions, paddle configuration and dimensions, velocity gradient, retention time, and power requirement.

(5.4) The expansion of granular filter beds under backwashing is governed by Equation 5.21. Write a program which will tabulate the expansion of a bed containing single-size particles for a range of backwash velocities. Develop this program to calculate expansion for a stratified bed made up of five different particle size layers.

FURTHER READING

Institute of Water Pollution Control, *Primary Sedimentation*, IWPC, 1980.

Tebbutt, T. H. Y., *Principles of Water Quality Control*, 3rd edn, Chapters 9, 10, 11 and 13, Pergamon Press, 1983.

Chapter 6

Chemical treatment processes

As outlined in the previous chapter, chemical treatment processes involve the use of chemical reactions to assist in the removal of impurities from waters and wastewaters. In some cases the process may be entirely chemical in nature, whereas in others the chemical process may be used in combination with a physical process to achieve the desired result.

ESSENTIAL THEORY

6.1 Chemical coagulation

The settlement of fine colloidal and suspended solids can be assisted by the use of flocculation as described in Section 5.2 but with dilute suspensions, such as lowland river waters, the opportunities for collisions and agglomeration are limited. In such circumstances flocculation does not significantly improve the settling characteristics of the suspension. The addition of a chemical coagulant, which precipitates flocculant solids in the water, followed by flocculation and sedimentation, can provide a high degree of clarification. The process of chemical coagulation is thus carried out in a sequence of operations, the first of which involves rapid mixing of the coagulant with the flow of water to permit enmeshment of the colloidal solids in the rapidly precipitating floc. This can be achieved by means of hydraulic turbulence or by a high-speed rotating-blade mixer with a retention time of 30–60 s. Following the initial precipitation stage, the suspension is passed to flocculation and sedimentation units, which may be separate or combined.

The most popular coagulant for potable water treatment is aluminium sulphate, often referred to as alum. When it is added to water in small doses of around 20–50 mg/l, a reaction takes place with the natural alkalinity present, and insoluble aluminium hydroxide is formed. This is a highly flocculent precipitate which responds well to controlled flocculation. The chemical reactions which occur are complex but may be simplified as

$$Al_2(SO_4)_3 + 3Ca(HCO_3)_2 \rightarrow 2Al(OH)_3 + 3CaSO_4 + 6CO_2$$

When using commercial alum which has 16–18 molecules of water of crystallization, and expressing alkalinity in terms of calcium carbonate, each mg/l of alum reacts with 0.5 mg/l of alkalinity. If insufficient alkalinity is present, the precipitation of aluminium hydroxide will not be complete, coagulation will be poor and residual aluminium will appear in the treated water. The solubility of aluminium hydroxide is lowest in the pH range of 5–7.5, and outside this range coagulation with alum will not be satisfactory. An acid or alkali, as appropriate, can be added to bring the pH into the correct range. Alternatively, another coagulant such as ferric sulphate, ferrous chloride or ferric sulphate may be employed. The size and strength of floc particles can often be enhanced by the use of a small dose ($\ll$ 1 mg/l) of a coagulant aid such as a polyelectrolyte. These compounds are very large organic molecules which serve to bind floc particles together and thus improve their settling characteristics and resistance to shear.

There is no theoretical basis for the calculation of coagulant doses and they must be determined experimentally using a jar-test apparatus which simulates the stages of rapid mixing, flocculation and sedimentation. By using a range of coagulant doses and pH values, it is possible to determine the optimum conditions to achieve the required water quality.

6.2 Chemical precipitation

In essence, chemical precipitation depends upon the use of an added reagent which combines with the impurity to be removed to give an insoluble product which can then be removed by sedimentation, preceded by flocculation if necessary.

$$\underset{\text{impurity}}{A} + \underset{\text{reagent}}{B} \rightarrow \underset{\text{precipitate}}{C} + \underset{\text{byproduct}}{D}$$

It is clearly essential that any byproduct of the reaction does not itself have undesirable properties in relation to the eventual use of the water or wastewater. It is also important to remember that chemical precipitation processes produce sludges containing the impurities and that the cost of handling and disposing of these sludges in a safe manner can be significant.

To illustrate the way in which chemical precipitation may be used, it is convenient to consider the softening of water. Hardness in water is due to the presence of calcium and/or magnesium which may be associated with carbonates, bicarbonates, sulphates and chlorides. The metallic cations actually cause the effect of hardness (scale formation in hot water systems and reduced efficiency of soap) but the anions with which they are associated control the behaviour of the hardness. It is perhaps worth noting that although there are often economic reasons for reducing hardness, there is evidence to suggest that hard waters have a beneficial effect in relation to some heart diseases. Many natural waters contain material which has entered by dissolution of insoluble rocks under the action of water and biologically produced carbon dioxide:

$$\underset{\text{insoluble}}{CaCO_3} + H_2O + CO_2 \rightarrow \underset{\text{soluble}}{Ca(HCO_3)_2}$$

and reversal of this process can be used to remove the hardness.

A convenient way to set out the steps in chemical precipitation softening is to use a bar diagram which expresses the composition of the water in terms of calcium carbonate. This is obtained from the water analysis by multiplying the concentration of each constituent by the ratio:

equivalent weight of $CaCO_3$/equivalent weight of constituent

Thus

55 mg/l Ca^{2+} $= 55 \times (100/2)/(40/2) = 137.5$ mg/l as $CaCO_3$

10 mg/l Na^+ $= 10 \times (100/2)/23 = 21.7$ mg/l as $CaCO_3$

125 mg/l HCO_3^- $= 125 \times (100/2)/61 = 102.5$ mg/l as $CaCO_3$

40.2 mg/l Cl^- $= 40.2 \times (100/2)/35.5 = 56.7$ mg/l as $CaCO_3$

Figure 6.1 shows how the composition of this water may be represented as a block diagram; note that when expressed in terms of calcium carbonate, the concentrations of cations and anions are equal.

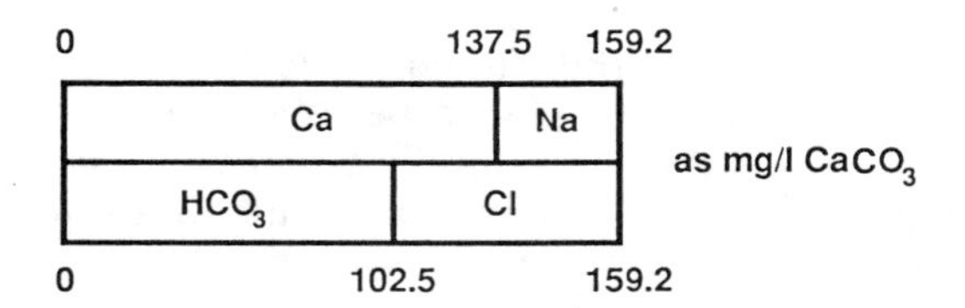

Figure 6.1 Block diagram for composition of water

Lime softening utilizes the reaction which occurs when lime is added to a water containing calcium hardness associated with bicarbonates:

$$Ca(HCO_3)_2 + Ca(OH)_2 \rightarrow 2CaCO_3 + 2H_2O$$

In practice, the solubility of calcium carbonate is about 40 mg/l, so the hardness cannot be removed below this level by precipitation. The steps in lime softening are shown in Figure 6.2.

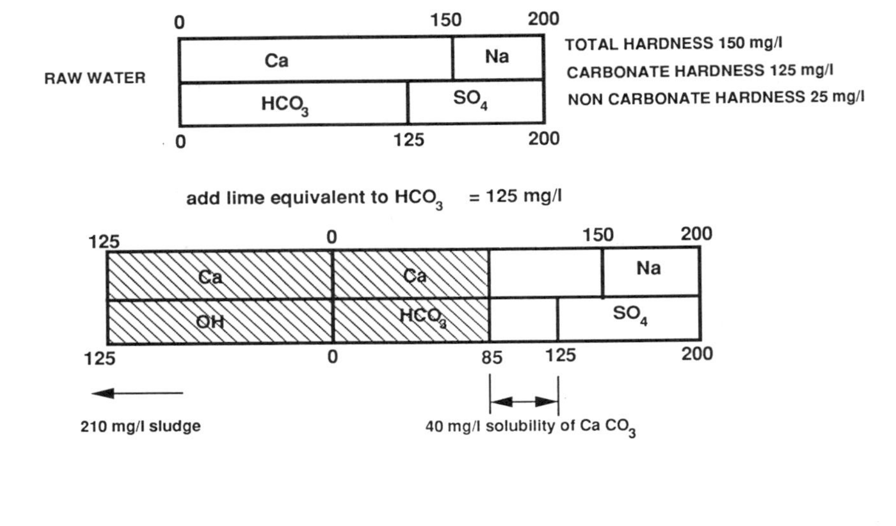

Figure 6.2 Lime softening

If calcium hardness of both carbonate and non-carbonate forms is present, the lime soda process must be used as shown in Figure 6.3. Here the first stage is as for lime softening but this is followed by the addition of sodium carbonate to allow further precipitation of calcium carbonate:

$$CaSO_4 + Na_2CO_3 \rightarrow CaCO_3 + Na_2SO_4$$

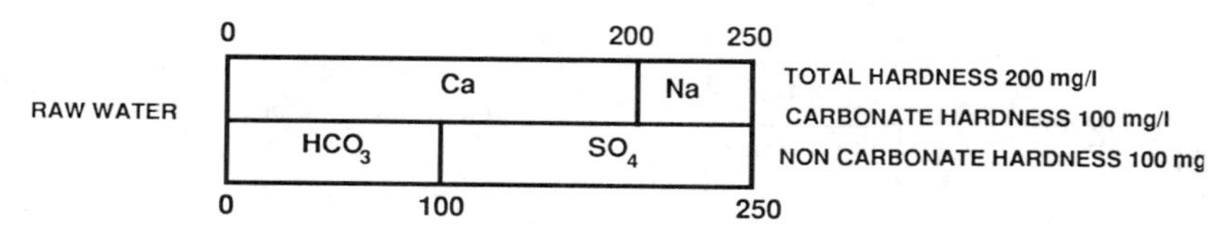

add lime equivalent to HCO_3 = 100 mg/l

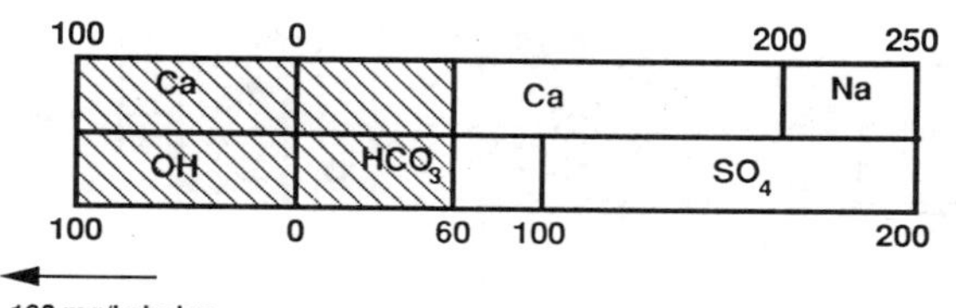

add sodium carbonate equivalent to non carbonate hardness associated with Ca = 100 mg/l

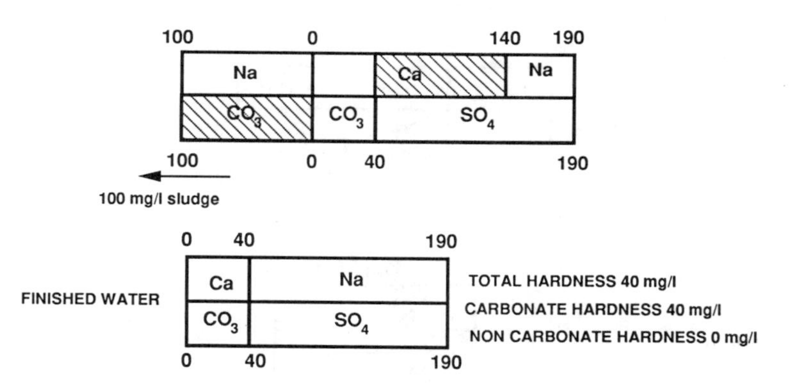

Figure 6.3 Lime soda softening

When magnesium hardness is present, the softening process is complicated because magnesium carbonate is soluble:

$$Mg(HCO_3)_2 + Ca(OH)_2 \rightarrow CaCO_3 + MgCO_3 + 2H_2O$$

However, by raising the pH to 11, magnesium hydroxide is precipitated with a solubility of 10 mg/l:

$$MgCO_3 + Ca(OH)_2 \rightarrow Mg(OH)_2 + CaCO_3$$

Figure 6.4 illustrates the excess lime process used for this type of hardness. It will be noted that the process is fairly complicated and that it produces large volumes of sludge. When both

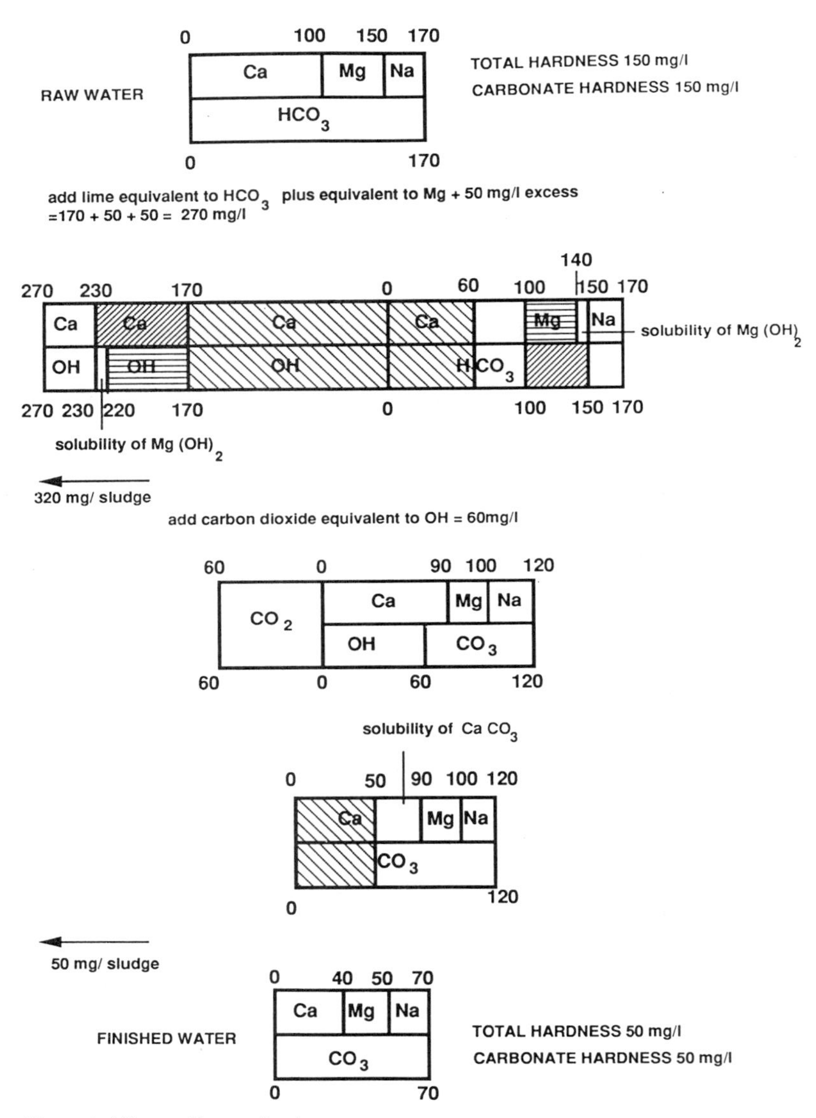

Figure 6.4 Excess lime softening

calcium and magnesium carbonate and non-carbonate forms of hardness are present, the even more complex excess lime soda process can be used, although ion-exchange treatment may be more economic in such cases. Softened waters should be stabilized to prevent further precipitation of calcium carbonate scale by the addition of carbon dioxide, which converts the carbonate back into soluble bicarbonate, or of polyphosphates, which keep carbonate in suspension as a floc. This stabilization is important, since otherwise the distribution system carrying the softened water can have its capacity seriously reduced by calcium carbonate scale.

Another application of chemical precipitation is for the treatment of industrial wastewaters containing metals such as chromium. These wastewaters arise from metal-finishing operations like chromium plating and can cause serious damage to sewage treatment processes if not subjected to effective pretreatment before discharge to the sewer. Chromic acid and chromate wastes can be treated by reduction of the chromate using ferrous sulphate and adding lime to cause precipitation of chromium as the hydroxide. This can then be removed by sedimentation. The basic equations for the reactions when the chromium is present in acid solution, which is usually the case, are:

$$2CrO_3 + 6FeSO_4.7H_2O + 6H_2SO_4 \rightarrow$$

$$3Fe_2(SO_4)_3 + Cr_2(SO_4)_3 + 13H_2O + 12Ca(OH)_2 \rightarrow$$

$$6Fe(OH)_3 + 2Cr(OH)_3 + 12CaSO_4$$

(all precipitated as sludge)

6.3 Ion exchange

A number of natural materials such as zeolites, which are complex sodium–aluminium silicates, have the ability to exchange ions in their structure for other ions with which they come into contact. Synthetic ion-exchange resins have been developed to provide higher exchange capacities or for specialized applications. Exchange resins provide either cation exchange (positive ions) or anion exchange (negative ions) and the commonest application in water treatment is as an alternative to precipitation methods in softening. Ion-exchange softening uses a sodium-cycle cation exchange which takes up calcium and magnesium ions in the water passing through the bed and releases sodium ions back into the water. This gives complete removal of hardness, since sodium does not contribute to hardness. The presence of high concentrations of sodium in water is, however, undesirable for people suffering from some heart complaints.

The ion-exchange reaction can be depicted as:

$$\begin{matrix} Ca^{2+} \\ Mg^{2+} \end{matrix} \left\{ \begin{matrix} HCO_3^{-} \\ CO_3^{2-} \\ SO_4^{2-} \\ Cl^{-} \end{matrix} \right. + Na_2X \rightarrow \left. \begin{matrix} NaHCO_3 \\ Na_2CO_3 \\ Na_2SO_4 \\ NaCl \end{matrix} \right\} + \begin{matrix} CaX \\ MgX \end{matrix}$$

When all the available sodium has been released, no further softening can take place but the ion-exchange material can then be regenerated by bringing it into contact with a strong solution of sodium, for which common salt solution is the most readily available form:

$$\begin{matrix} CaX \\ MgX \end{matrix} + 2NaCl \rightarrow Na_2X + \begin{matrix} CaCl_2 \\ MgCl_2 \end{matrix}$$

The calcium and magnesium ions appear as a waste stream of chlorides which must be disposed of in a suitable manner; the ion-exchange material can be used again and again.

The performance of an ion exchanger is expressed in terms of its exchange capacity (g equivalents removed/m^3) and its regenerant requirement (equivalents/equivalent exchanged). For a natural zeolite these factors would be of the order of 200 g/m^3 and 5 equiv./equiv., respectively.

Ion exchange can also provide a means for removing nitrates from water using a cation-exchange system, although at the present time nitrate-specific resins are not fully developed. Ion-exchange methods can also be used to treat certain industrial wastewaters, particularly those containing metals in solution. The recovery for re-use of silver from photographic processing wastes is a good example of the use of an ion-exchange technique for both pollution control and conservation of raw materials. It must be noted that, in common with most treatment processes, ion-exchange methods do not destroy the contaminants but merely make it possible to remove and concentrate them. The final disposal or re-use of the contaminants needs careful consideration.

6.4 Disinfection

There are well-established links between the contamination of drinking water with faecal matter and the incidence of such water-related diseases as cholera, typhoid and many gastro-intestinal infections. Thus the removal of the pathogenic microorganisms from water supplies is a very valuable measure for the improvement of public health. Because of the small size of the bacteria, which are responsible for many waterborne diseases, it is not possible to guarantee that all will be removed by coagulation and filtration processes, and their absence from groundwaters cannot be assumed. It is therefore standard practice to disinfect water before distribution. Disinfection means the destruction of pathogenic microorganisms and does not necessarily mean that the water is sterile, since a small number of harmless microorganisms is usually present in tap water and poses no hazard.

For most disinfectants the rate of kill is given by

$$dN/dt = -KN \tag{6.1}$$

where K = rate constant
N = number of living microorganisms

The rate constant varies with the particular disinfectant, its concentration, the organism being killed, pH, temperature, and other environmental factors.

The most common disinfectant in water treatment is chlorine, which is available in a number of forms and which reacts in different ways when added to water. In the absence of ammonia, free chlorine residuals are formed, but reactions with ammonia form combined residuals which are less effective disinfectants than free residuals. Chlorine has the advantage of being a powerful disinfectant which is cheap and readily available but its reactions with trace organic compounds to form potentially carcinogenic organochlorine compounds cause some concern. Other disinfectants which can be used in water treatment include ozone and ultraviolet radiation but these also have some disadvantages.

The behaviour of chlorine and its numerous compounds in water is complex and thus its disinfecting action is not appropriately described by Equation 6.1. A better representation is usually given by:

$$dN/dt = -KNt \tag{6.2}$$

where t = time of exposure

Integrating and changing to base 10 gives

$$t^2 = (2/k) \log (N_0/N_t) \tag{6.3}$$

where N_0 = number of microorganisms initially
N_t = number of organisms remaining after time t

Because of the way in which the disinfecting action of chlorine depends upon several factors, another approach to modelling the process has resulted in the following expression:

$$S_L c^n t = -\log(N_0/N_t) \quad (6.4)$$

where S_L = coefficient of specific lethality
c = concentration of disinfectant
n = dilution coefficient

For a particular microorganism and a given percentage kill it is then possible to write

$$c^n t = \text{constant} \quad (6.5)$$

For a free chlorine residual, a value of $n = 0.86$ is often used.

WORKED EXAMPLES

Example 6.1 COAG: chemical coagulation

Write a program to calculate the weight of chemical required to give coagulation of a water. The user should be able to input details of the flow and the type of coagulant to be used.

At a river abstraction plant a flow of 0.1 m^3/s is to be given chemical coagulation using either aluminium sulphate, ferrous sulphate or ferric chloride. The coagulant doses obtained from jar tests are expressed in terms of aluminium and iron.
(Atomic weights: Al 27, Cl 35.5, Fe 56, H 1, O 16, S 32)
(Commercial alum $Al_2(SO_4)_3.14H_2O$)
(Anhydrous ferrous sulphate $Fe_2(SO_4)_3$)
(Crystalline ferric chloride $FeCl_3$)

```
10 CLS
20 REM COAG
30 PRINT"Chemical Coagulation"
40 INPUT"Enter flow to be treated (cu.m/s)";Q
50 INPUT"Enter coagulant dose (mg/l as Al or Fe)";D
60 INPUT"Enter A for aluminium or F for iron";C$
70 IF C$="A" OR C$="a" THEN 100
80 IF C$="F" OR C$="f" THEN 150
90 PRINT"Enter A or F only!":GOTO 60
100 CD=D*(2*27+3*(32+4*16)+14*(2*1+16))/(2*27)
110 GOSUB 300
120 PRINT
130 PRINT"Commercial aluminium sulphate addition=";CW;"kg/d"
140 GOTO 250
150 CD=D*(2*56+3*(32+4*16))/(2*56)
160 GOSUB 300
170 PRINT
180 PRINT"Anhydrous ferrous sulphate addition=";CW;"kg/d"
190 CD=D*(56+3*35.5)/56
```

```
200 GOSUB 300
210 PRINT
220 PRINT"or alternatively -"
230 PRINT
240 PRINT"Crystalline ferric chloride addition=";CW;"kg/d"
250 PRINT
260 INPUT"Another calculation (Y/N)";Q$
270 IF Q$="Y" OR Q$="y" THEN 10
280 END
300 CW=CD*Q*60*60*24*.001
310 RETURN
```

```
Chemical Coagulation
Enter flow to be treated (cu.m/s)? .1
Enter coagulant dose (mg/l as Al or Fe)? 7.5
Enter A for aluminium or F for iron? A

Commercial aluminium sulphate addition= 712.8001 kg/d

Another calculation (Y/N)? Y

Chemical Coagulation
Enter flow to be treated (cu.m/s)? .1
Enter coagulant dose (mg/l as Al or Fe)? 7.5
Enter A for aluminium or F for iron? F

Anhydrous ferrous sulphate addition= 231.4286 kg/d

or alternatively -

Crystalline ferric chloride addition= 188.0357 kg/d

Another calculation (Y/N)?
```

Program notes

(1) Lines 10–30 set up title
(2) Lines 40–60 input information about flow and coagulant
(3) Lines 70–80 direct program to appropriate coagulant
(4) Line 90 error traps for wrong response in 60
(5) Line 100 calculates commercial aluminium sulphate dose corresponding to Al dose
(6) Line 110 directs program to subroutine for calculation of daily weight of chemical
(7) Lines 120–130 print answer
(8) Line 140 directs program to another calculation query
(9) Line 150 calculates ferrous sulphate dose
(10) Line 160 directs program to daily weight subroutine
(11) Lines 170–180 print answer
(12) Lines 190–240 calculate alternative ferric chloride usage
(13) Lines 250–270 offer another calculation if desired

Example 6.2 PRESOFT: precipitation softening

Develop a program which allows the user to determine the effect of softening a water by various precipitation techniques. It should allow input of analytical data in the form of individual constituents or as calcium carbonate. Results should give the hardness of the finished water, chemical needs and sludge production.
(Atomic weights: C 12, Ca 40.08, H 1, Mg 24.32, O 16)

```
10 CLS
20 REM PRESOFT
30 PRINT"Precipitation Softening"
40 INPUT"Analyses already in terms of CaCO3 (Y/N)";Q$
50 IF Q$="Y" OR Q$="y" THEN 140
60 INPUT"Enter mg/l of cations: Ca, Mg";C,M
70 INPUT"Enter mg/l of HCO3,";H
80 CA=C*50/20.04:MG=M*50/12.16:HC=H*50/61
90 PRINT
100 IF CA<>0 THEN PRINT"Ca as mg/l CaCO3=";CA
110 IF MG<>0 THEN PRINT"Mg as mg/l CaCO3=";MG
120 IF HC<>0 THEN PRINT"HCO3 as mg/l CaCO3=";HC
130 GOTO 160
140 INPUT"Enter Ca, and Mg (mg/l as CaCO3)";CA,MG
150 INPUT"Enter HCO3 (mg/l as CaCO3)";HC
160 IH=CA+MG
170 IF IH>HC THEN CH=HC
180 IF IH<=HC THEN CH=IH
190 NC=IH-HC
200 IF IH=HC THEN NC=0
210 C$="mg/l as CaCO3"
220 PRINT"Total hardness =";IH C$
230 PRINT"Carbonate hardness =";CH C$
240 PRINT"Non carbonate hardness =";NC; C$
250 IF CA<=40 THEN 460
260 PRINT
270 L=HC
280 S1=L+HC-40: IF S1<0 THEN S1=0
290 TH=IH-HC+40:IF HC=0 THEN TH=IH
300 IF HC>CA THEN S1=L+CA-40
310 IF HC>CA THEN TH=IH-CA+40
320 PRINT"After lime softening:"
330 PRINT"Total hardness =";TH; C$
340 PRINT"Lime dose =";L; C$
350 PRINT"Sludge produced =";S1; C$
360 IF TH<=40 THEN 700
370 PRINT
380 SC=CA-CH:IF SC<=0 THEN 470
390 S2=S1+SC:IF S1=0 THEN S2=SC-40
400 TH=IH-CA+40
410 PRINT"After lime-soda softening:"
420 PRINT"Total hardness =";TH C$
430 PRINT"Lime dose =";L C$
440 PRINT"Sodium carbonate dose =";SC C$
450 PRINT"Sludge produced =";S2; C$
460 IF TH=40 THEN 700
470 IF HC<IH THEN 570
480 PRINT
490 L=HC+MG+50
500 S3=HC+(CA-40)+MG+(MG-10)+50
510 TH=IH-(CA-40)-(MG-10)
```

```
520 PRINT"After excess lime softening:"
530 PRINT"Total hardness =";TH C$
540 PRINT"Lime dose =";L C$
550 PRINT"Carbon dioxide dose = 60";C$
560 PRINT"Sludge produced =";S3 C$
570 IF TH<=50 THEN 700
580 PRINT
590 L=HC+MG+50
600 SC=NC-10
610 S4=HC+(HC-40)+(MG-10)+60+SC
620 IF HC>CA THEN S4=HC+(CA-40)+(MG-10)+(HC-CA)+60+SC
630 TH=IH-(CA-40)-(MG-10)
640 PRINT"After excess lime soda softening:"
650 PRINT"Total hardness =";TH C$
660 PRINT"Lime dose =";L C$
670 PRINT"Sodium carbonate dose =";SC C$
680 PRINT"Carbon dioxide dose = 60";C$
690 PRINT"Sludge produced =";S4 C$
700 PRINT
710 PRINT"Minimum hardness by precipitation method"
720 PRINT
730 INPUT"Another set of analyses (Y/N)";A$
740 IF A$="Y" OR A$="y" THEN 10
750 END
```

```
Precipitation Softening
Analyses already in terms of CaCO3 (Y/N)? Y
Enter Ca, and Mg (mg/l as CaCO3)? 150,0
Enter HCO3 (mg/l as CaCO3)? 125
Total hardness = 150 mg/l as CaCO3
Carbonate hardness = 125 mg/l as CaCO3
Non carbonate hardness = 25 mg/l as CaCO3

After lime softening:
Total hardness = 65 mg/l as CaCO3
Lime dose = 125 mg/l as CaCO3
Sludge produced = 210 mg/l as CaCO3

After lime-soda softening:
Total hardness = 40 mg/l as CaCO3
Lime dose = 125 mg/l as CaCO3
Sodium carbonate dose = 25 mg/l as CaCO3
Sludge produced = 235 mg/l as CaCO3

Minimum hardness by precipitation method

Another set of analyses (Y/N)? Y


Precipitation Softening
Analyses already in terms of CaCO3 (Y/N)? Y
Enter Ca, and Mg (mg/l as CaCO3)? 100,50
Enter HCO3 (mg/l as CaCO3)? 170
Total hardness = 150 mg/l as CaCO3
Carbonate hardness = 150 mg/l as CaCO3
Non carbonate hardness =-20 mg/l as CaCO3

After lime softening:
Total hardness = 90 mg/l as CaCO3
Lime dose = 170 mg/l as CaCO3
Sludge produced = 230 mg/l as CaCO3


After excess lime softening:
Total hardness = 50 mg/l as CaCO3
```

```
Lime dose = 270 mg/l as CaCO3
Carbon dioxide dose = 60mg/l as CaCO3
Sludge produced = 370 mg/l as CaCO3

Minimum hardness by precipitation method

Another set of analyses (Y/N)? N
```

Program notes

(1) Lines 10–30 set up title
(2) Line 40 queries form of analytical data
(3) Line 50 directs program to 140 if analyses already in terms of calcium carbonate
(4) Lines 60–80 input and manipulate analytical data
(5) Lines 90–120 print composition as calcium carbonate
(6) Line 130 directs program to skip alternative data input
(7) Lines 140–150 input analytical data as calcium carbonate
(8) Lines 160–200 calculate hardness values
(9) Line 210 string variable for mg/l $CaCO_3$
(10) Lines 220–240 print hardness values
(11) Line 250 no removable calcium present
(12) Lines 270–350 calculate and print lime softening
(13) Line 360 no further softening possible
(14) Lines 380–450 calculate and print lime soda softening
(15) Line 460 no further softening possible
(16) Line 470 directs program to excess lime soda
(17) Lines 490–560 calculate and print excess lime softening
(18) Line 570 no further softening possible
(19) Lines 590–690 calculate and print excess lime soda
(20) Lines 700–750 end routines

Example 6.3 IONEX: Sodium cycle ion-exchange softening

Write a program which will calculate the performance of a sodium cycle ion exchanger used for softening a water whose analysis is provided by the user. The user should be able to enter the exchange capacity and regenerant requirement of the resin.

A sodium cycle cation exchanger has an exchange capacity of 400 g equivalents/m^3 and a regeneration requirement of 5 equiv./equiv. exchanged. Determine the volume of water containing 50 mg/l of calcium and 20 mg/l of magnesium which can be softened per m^3 of exchanger and the regenerant requirement of sodium chloride per m^3 of water softened. (Atomic weights: C 12, Ca 40.08, Cl 35.5, Na 23, O 16)

```
10 CLS
20 REM IONEX
30 PRINT"Sodium Cycle Ion Exchange Softening"
40 INPUT"Are analyses already in terms of CaCO3 (Y/N)"; Q$
50 IF Q$="Y" OR Q$="y" THEN 120
60 INPUT"Enter mg/l of cations: Ca, Mg";C,M
70 CA=C*50/20.04:MG=M*50/12.16
80 PRINT
90 IF CA<>0 THEN PRINT"Ca as mg/l CaCO3=";CA
100 IF MG<>0  THEN PRINT"Mg as mg/l CaCO3=";MG
110 GOTO 130
120 INPUT"Enter Ca, and Mg (mg/l as CaCO3)";CA,MG
130 IH=CA+MG
140 INPUT"Enter resin exchange capacity (g equiv/cu.m)";XC
150 INPUT"Enter regenerant requirement (equiv/equiv)";RR
160 GE=IH/50
170 VT=XC/GE
180 PRINT"Volume of water softened/cu.m of resin=";VT;"cu.m"
190 SR=GE*RR*58.5/1000
200 PRINT"Salt regenerant=";SR;"kg/cu.m water softened"
210 PRINT
220 INPUT"Another calculation (Y/N)";A$
230 IF A$="Y" OR A$="y" THEN 40
240 END
```

```
Sodium Cycle Ion Exchange Softening
Are analyses already in terms of CaCO3 (Y/N)? N
Enter mg/l of cations: Ca, Mg? 50,20

Ca as mg/l CaCO3= 124.7505
Mg as mg/l CaCO3= 82.23685
Enter resin exchange capacity (g equiv/cu.m)? 400
Enter regenerant requirement (equiv/equiv)? 5
Volume of water softened/cu.m of resin= 96.62427 cu.m
Salt regenerant= 1.210876 kg/cu.m water softened

Another calculation (Y/N)? N
```

Program notes

(1) Lines 10–30 set up title
(2) Line 40 queries form of analytical data
(3) Line 50 directs program to 120 if analyses already in form of calcium carbonate
(4) Lines 60–100 input and manipulate analytical data
(5) Line 110 directs program to skip alternative data input
(6) Line 120 inputs analytical data as calcium carbonate
(7) Line 130 calculates initial total hardness
(8) Lines 140–150 input resin characteristics
(9) Line 160 calculates gram equivalent of hardness
(10) Lines 170–180 calculate and print volume treated
(11) Lines 190–200 calculate and print regenerant required
(12) Lines 210–230 offer another set of data if desired

Example 6.4 KILL: disinfection calculations

A program is needed to print a table of contact time in minutes and free chlorine concentration in mg/l required for a 99% kill of various microorganisms. It can be assumed that the relevant relationship is given by

$$c^{0.86}t = D$$

where $D = 0.24$ for *E. coli*, 1.2 for polio virus and 6.3 for Coxsackie virus.

```
10 CLS
20 REM KILL
30 PRINT"Disinfection Calculations"
40 PRINT"Disinfection rate depends upon type of microorganism"
50 PRINT"Enter number indicated for particular microorganism"
60 INPUT"E.coli 1, Polio virus 2, Coxsackie virus 3";S
70 ON S GOTO 80, 90 ,100
80 TC=.24:SP$="E.coli":GOTO 110
90 TC=1.2:SP$="Polio virus":GOTO 110
100 TC=6.3:SP$="Coxsackie virus"
110 CLS
120 PRINT"Disinfection of ";SP$;"  free chlorine, 99% kill"
130 PRINT
140 PRINT"Contact time","Concentrn"
150 PRINT"   (min)    ","  (mg/l) "
160 FOR T= 1 TO 15
170 C=(TC/T)^(1/.86)
180 PRINT T,C
190 NEXT T
200 PRINT
210 INPUT"Another microorganism (Y/N)";Q$
220 IF Q$="Y" OR Q$="y" THEN 50
230 END
```

```
Disinfection Calculations
Disinfection rate depends upon type of microorganism
Enter number indicated for particular microorganism
E.coli 1, Polio virus 2, Coxsackie virus 3? 2

Disinfection of Polio virus  free chlorine, 99% kill

Contact time  Concentrn
   (min)        (mg/l)
 1             1.23615
 2             .5521236
 3             .3445713
 4             .2466048
 5             .1902459
 6             .153902
 7             .1286468
 8             .1101455
 9             9.604769E-02
 10            8.497293E-02
 11            7.605884E-02
 12            6.873998E-02
 13            6.263086E-02
 14            5.745983E-02
 15            5.303022E-02

Another microorganism (Y/N)? n
```

Program notes

(1) Lines 10–30 set up title
(2) Lines 40–60 specify type of microorganism
(3) Line 70 sends program to appropriate constant
(4) Lines 80–100 constants for specific microorganisms
(5) Lines 110–150 print results table
(6) Line 160 starts calculation loop
(7) Lines 170–180 calculates and prints result
(8) Line 190 returns program to start of loop
(9) Lines 200–220 offer another organism if desired

PROBLEMS

(6.1) When using chemical coagulation to remove colour from soft upland catchment waters there may be insufficient alkalinity in the raw water to permit the coagulation reaction to proceed efficiently. Using the fact that 1 mg/l of commercial aluminium sulphate destroys 0.5 mg/l of alkalinity as calcium carbonate, write a program which checks for the presence of sufficient alkalinity when the aluminium dose is entered and gives the user the option of specifying a minimum residual alkalinity in the treated water. It should also calculate the weights of lime or sodium carbonate needed to make up any deficiency in the alkalinity.
(Atomic weights: C 12, Ca 40, H 1, O 16, Na 23)

(6.2) Write a program to calculate the amounts of ferrous sulphate and lime necessary to precipitate hexavalent chromium from an industrial waste.
(Atomic weights: Ca 40, Cr 52, Fe 56, H 1, O 16, S 32)

(6.3) Using Worked Examples 6.2 and 6.3 as a basis, develop a program which will calculate the chemical costs of softening a water by precipitation and ion-exchange techniques and indicate the more economical process in terms of operating costs. Remember that precipitation softening does not produce zero-hardness water whereas ion-exchange treatment will remove all the hardness. This difference could be allowed for by bypassing a portion of the flow around the ion-exchange unit so that the final blended water hardness would be the same as with precipitation treatment.
(Relative costs of lime, sodium carbonate and sodium chloride for commercial grades are: 6 : 4 : 3)

(6.4) Using the relationship given in Equation 6.3 for disinfection with chlorine compounds, write a program which tabulates the contact times necessary to give kills of 99, 99.9, 99.99 and 99.999% of *E. coli* with both free and combined residuals. (Typical values of k (base 10) in Equation 6.3 are 10^{-2}/s for free chlorine and 10^{-5}/s for combined chlorine)

FURTHER READING

Tebbutt, T. H. Y., *Principles of Water Quality Control*, 3rd edn, Chapters 12, 16 and 17, Pergamon Press, 1983.

Chapter 7

Biological treatment processes

In many wastewaters the main pollutants are organic compounds in colloidal and soluble forms which are usually not readily removed by physical or chemical processes. Most organic substances can, however, be stabilized by microorganisms, so that biological treatment is a useful technique for wastewaters like domestic sewage, food-processing wastes, etc. It is not so appropriate for dealing with very low concentrations of organic matter.

ESSENTIAL THEORY

7.1 Principles of biological treatment

There are two main types of biological reaction, depending upon the presence or absence of free oxygen, as shown in Figure 4.2. The aerobic reaction takes place only in the presence of free oxygen and produces stable, relatively inert end products. Anaerobic reactions are more complex, being two-stage, proceeding relatively slowly and leading to end products which are unstable and which still contain considerable amounts of energy. In any biological reaction the energy in the organic matter, used as food by the microorganisms, is split three ways; some is used in creating new microorganisms, some is incorporated in the end products of the reaction, and the remainder appears as heat. The proportions of energy in the three areas depend upon the nature of the reaction, the type of organic matter, the type of microorganism, and environmental conditions. The organic matter in wastewater thus provides the basic requirements for the formation of new microorganisms as well as providing the energy for the oxidation reactions which release the end products. During biological oxidation, microbial cells are continually dying and the dead cells are broken down by auto-digestion, usually referred to as endogenous respiration, as illustrated in Figure 7.1.

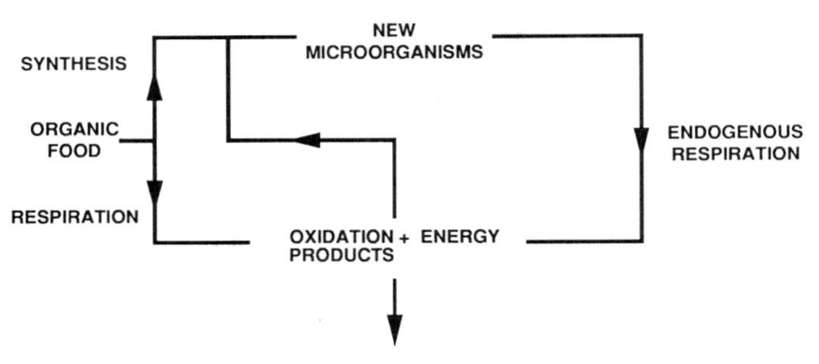

Figure 7.1 Synthesis, energy release and endogenous respiration

The classical biological growth curve, Figure 7.2, assumes:

1. That ample supplies of carbon, nitrogen and phosphorus are present: the empirical chemical formula for a bacterial cell is $C_{60}H_{87}O_{23}N_{12}P$ and in practice 1 part of organic and ammonia nitrogen is required for 15–30 parts of BOD, and 1 part of phosphorus for each 80–150 parts of BOD, depending upon the type of treatment in use.
2. Sufficient energy in the organic matter to support the reaction.
3. Sufficient inorganic ions: calcium, magnesium, potassium, iron, manganese, cobalt, etc.
4. Any necessary growth factors, such as vitamins, are available.

Domestic sewage satisfies all these requirements but some industrial wastewaters may be lacking in nutrients or other factors, which could result in inhibition of biological activity. Toxic substances present in the wastewater can also mean that biological treatment of an organic wastewater is ineffective, although it is sometimes possible for microorganisms to become acclimated to substances which initially appear to be toxic. The aim of conventional biological treatment processes is to achieve almost complete removal of the organic matter in the feed. As can be seen from Figure 7.2, this implies a relatively long retention time and a large volume of new cells which appears as sludge in biological treatment plants.

The rate of biochemical oxidation reactions depends upon temperature but cannot otherwise be significantly altered. However, by utilizing the adsorptive properties of biological materials it is possible to carry out biological treatment in a reactor with a relatively short hydraulic retention time. The

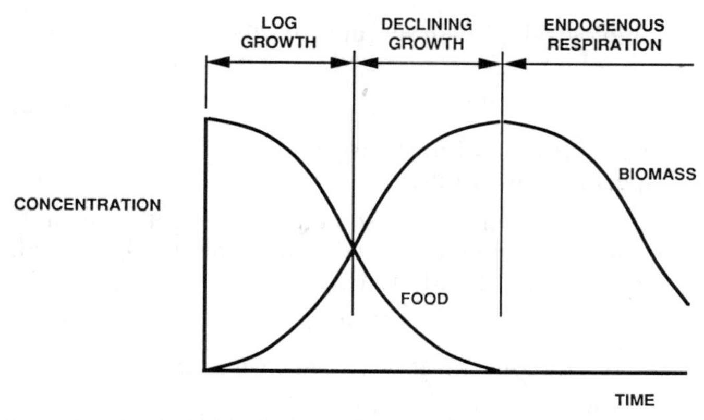

Figure 7.2 Classical biological growth curve

organic matter is initially adsorbed onto the biological solids and then broken down. It is thus essential that the reactor is able to provide the environment for growth and retention of large numbers of microorganisms. In fixed-film systems the biological growth occurs as a slime or film on the surfaces of a suitable medium such as stone or plastic; the biological filter is the commonest form of fixed-film reactor. Dispersed-growth systems maintain a high population of microorganisms in suspension by providing agitation through the introduction of gas (air for aerobic systems) or by mechanical mixing. The activated-sludge system is a common form of dispersed-growth reactor. Both systems require a settling facility to remove the excess biological solids produced in the process. In the case of fixed-film systems the solids are essentially dead cells, but with dispersed-growth activated-sludge systems the bulk of the cells are living and are returned to the reactor for re-use, only the excess solids being removed.

7.2 Biochemical reactions

At high concentrations of organic matter the rate of removal of organics is independent of its concentration, i.e.

$$dS/dt = K \qquad (7.1)$$

where S = concentration of organic matter (Chemical Organic Demand, COD or ultimate BOD)

t = time

K = rate constant

The rate of biological growth in unrestricted conditions in a batch reactor is

$$dX/dt = \mu_m X \tag{7.2}$$

where X = concentration of microorganisms
μ_m = specific growth rate

When growth becomes limited by some external factor, such as concentration of organic matter or reduced nutrients, the growth rate can be expressed by the Monod equation:

$$\mu = \mu_m S/(K_s + S) \tag{7.3}$$

where K_s = organic concentration at which $\mu = \mu_m/2$

The synthesis of new cells is given by

$$dX = YdS \tag{7.4}$$

where X = concentration of biological solids (volatile SS)
Y = yield coefficient (VSS produced/unit COD removed)

During any biological reaction, endogenous respiration takes place with a loss of cells:

$$\text{endogenous decay} = -bX \tag{7.5}$$

where b = endogenous decay coefficient

Thus the net rate of growth of biological solids is given by

$$Y(S_0 - S)/\theta - bX \tag{7.6}$$

where S_0 = initial organic concentration
θ = hydraulic retention time

From Equation 7.3, the rate of growth of biological solids is

$$\mu_m XS/(K_s + S) \tag{7.7}$$

and the rate of utilization of the organic matter must be

$$\mu_m XS/Y(K_s + S) \tag{7.8}$$

Hence

$$(S_0 - S)/\theta = \mu_m XS/Y(K_s + S) \tag{7.9}$$

This can be rewritten as

$$(S_0 - S)/\theta X = \mu_m S/Y(K_s + S) \tag{7.10}$$

This can be inverted and given a linear form:

$$X\theta/(S_0 - S) = K_s Y/\mu_m S + Y/\mu_m \tag{7.11}$$

In systems with solids recycle or fixed-film systems, the solids retention time (θ_c) may be much longer than the hydraulic retention time (θ):

$$\text{net growth of biological solids} = X/\theta_c \tag{7.12}$$

and hence

$$X/\theta_c = Y(S_0 - S)/\theta - bX \tag{7.13}$$

or

$$1/\theta_c = Y(S_0 - S)/\theta X - b \tag{7.14}$$

An aerobic reaction requires oxygen to oxidize the organic matter in the wastewater and also to support endogenous respiration. This oxygen requirement can be expressed as

$$O_2 = 0.39(S_0 - S) + 1.42bX \tag{7.15}$$

7.3 Biological filter

The biological filter or bacteria bed (Figure 7.3) is the simplest form of fixed-film system using a bed of stone or plastic medium. The process depends upon the formation of a film of microorganisms attached to the surface of the medium. The liquid waste flows past this film and air passes up through the bed to provide the oxygen which is essential for an aerobic reaction. If a strong waste is applied to a biological filter, excessive film growth will occur, the voids will become blocked and anaerobic conditions will be established with consequent deterioration in performance. To prevent this occurrence, strong wastewaters can be treated by recirculating the effluent round the filter. This serves the function of diluting the organic matter in the feed and prevents the ponding which is likely to result from the application of the concentrated wastewater. For small installations a specialized form of biological filter, the rotating biological contactor, provides space for film growth on discs which rotate partly submerged in the wastewater. These units

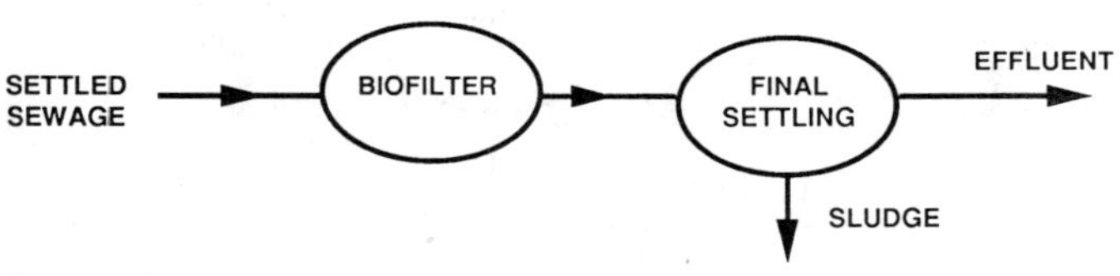

Figure 7.3 Biological filter flow sheet

are usually prefabricated and are provided with integral sedimentation chambers so that complete treatment is achieved in a single unit.

7.4 Activated sludge

In the activated-sludge process (Figure 7.4) a high concentration of biological solids (2000–5000 mg/l VSS) is kept in suspension as a floc by diffused air or mechanical aeration. The final settling tank separates out the solids as a sludge, the bulk of which is returned to the aeration tank to maintain the high solids concentration there. Activated-sludge systems depend for their successful performance on the settleability of the solids, and a number of control parameters are used to assist operation and control. An experimental measurement of sludge settleability is made by allowing a sample of the aeration tank mixed liquor to settle in a litre cylinder. The percentage of sludge after 30 minutes settlement is a good indication of the sludge separation properties and is used in the Sludge Volume Index (SVI) and Sludge Density Index (SDI).

$$\mathrm{SVI} = \text{sludge after 30 min } (\%) \,/\mathrm{MLSS}\ (\%) \qquad (7.16)$$

For good performance, SVI values of 40–100 are usual, but the value will exceed 200 for a sludge with poor settling characteristics, which will make it difficult to maintain the required MLSS concentration in the aeration unit.

$$\mathrm{SDI} = 100/\mathrm{SVI} \qquad (7.17)$$

The mean cell residence time (MCRT) in an activated-sludge system is given by

$$\theta_c = \frac{\text{aeration unit vol } (\mathrm{m}^3) \times \mathrm{MLVSS}\ (\mathrm{mg/l})}{\text{sludge wastage } (\mathrm{m}^3/\mathrm{day}) \times \text{sludge VSS } (\mathrm{mg/l})} \qquad (7.18)$$

Values of θ_c for conventional activated-sludge systems are

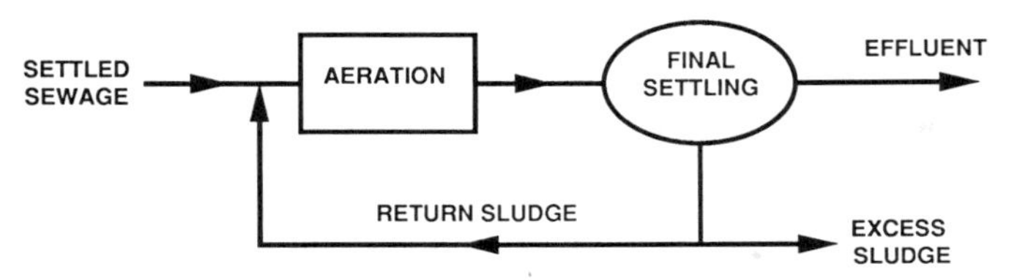

Figure 7.4 Activated-sludge flow sheet

usually around 3–4 days as compared with hydraulic retention times of a few hours.

7.5 Oxidation pond

In warm climates wastewater treatment can be effectively carried out using oxidation ponds, which are partly dispersed growth systems combining the activities of algae and bacteria. A symbiotic relationship occurs in which the algae utilize the inorganic end products resulting from bacteriological breakdown of organic matter and in turn produce oxygen by photosynthesis. This oxygen is then available for further aerobic oxidation of wastewater. The organic solids and dead microorganisms settle to the bottom of the pond where anaerobic reactions reduce their volume and produce some stabilization. Although ponds may be designed as aerobic or anaerobic systems, many are operated as facultative units with aerobic conditions prevailing but with anaerobic conditions existing in the lower regions of the pond for most of the time. An advantage of oxidation ponds is that as well as stabilizing organic matter they also provide considerable reductions in the number of pathogenic microorganisms. This is particularly important in developing countries where the pond effluent may be discharged to a stream or lake which is used for drinking water without treatment.

7.6 Anaerobic processes

With wastewaters having very high organic contents, e.g. food-processing wastes, or with the organic sludges from sedimentation of wastewaters, the oxygen demand may be so high that it becomes very difficult and expensive to maintain aerobic conditions. In such circumstances anaerobic processes can provide an efficient means of removing large amounts of organic matter, although they are not capable of producing very high quality effluents. The methane produced in anaerobic reactions can be a useful source of energy which can help to defray some of the costs of treatment. On a small scale, biogas units are widely used in some developing countries for waste treatment, energy supply and the production of agricultural fertilizer. The fundamental biochemical reactions outlined in Section 7.2 can be used for anaerobic systems and the methane production can be calculated from

$$CH_4 = 0.35[(S_0 - S) - 1.42 \times \text{VSS accumulation}] \quad (7.19)$$

WORKED EXAMPLES

Example 7.1 BIOGRO: solids accumulation in aerobic treatment

As shown in Equation 7.6, the accumulation of biological solids in a biological treatment system is a function of the combined effects of the yield coefficient and the endogenous respiration coefficient. In a system treating a wastewater, the amount of sludge produced can be determined by consideration of the VSS balance. This can be expressed as

VSS synthesized by removal of COD
+ any non-biodegradable VSS in the feed
− VSS in the effluent
− endogenous respiration VSS

It is thus possible to write a program to calculate the VSS accumulation, and also the oxygen requirement, for an activated-sludge system, for a range of retention times and operational conditions. The user provides influent and effluent COD and VSS, biodegradability factor for wastewater, and MLVSS for the reactor.

Determine the VSS accumulation and oxygen requirement in an aerobic system with MLVSS 2500 mg/l if the COD and VSS of the raw wastewater are 900 and 300 mg/l, the effluent is to be 40 mg/l COD and 20 mg/l VSS. Assume that the influent VSS are 95% biodegradable and that the appropriate values of the yield and endogenous respiration coefficients are 0.55 and 0.15 respectively.

```
10 CLS
20 REM BIOGRO
30 PRINT"Solids Accumulation in Aerobic Treatment"
40 INPUT"Enter initial COD, VSS of wastewater (mg/l)";C,V
50 INPUT"Enter biodegradability factor for wastewater";BF
60 INPUT"Enter effluent COD, VSS (mg/l)";CF,VF
70 INPUT"Enter MLVSS (mg/l)";ML
80 Y=.55:B=.15
90 CR=BF*(C-CF)
100 ST=V*(1-BF)
110 PRINT
120 PRINT"Retention"," Solids","Oxygen"
130 PRINT"  Time   "," Accuml"," Reqt "
140 PRINT"   (h)   ","(g/cu.m)","(g/cu.m)"
150 FOR H= 1 TO 16 STEP 2
160 SA=Y*CR+ST-VF-B*ML*H/24
170 OX=.39*CR+1.42*B*ML*H/24
180 PRINT H,SA,OX
190 NEXT H
200 PRINT
210 INPUT"Another calculation (Y/N)";Q$
220 IF Q$="Y" OR Q$="y" THEN 10
230 END
```

```
Solids Accumulation in Aerobic Treatment
Enter initial COD, VSS of wastewater (mg/l)? 900,300
Enter biodegradability factor for wastewater? .95
Enter effluent COD, VSS (mg/l)? 40,20
Enter MLVSS (mg/l)? 2500

Retention        Solids          Oxygen
  Time           Accuml           Reqt
   (h)           (g/cu.m)        (g/cu.m)
 1               428.725          340.8175
 3               397.475          385.1925
 5               366.225          429.5675
 7               334.975          473.9425
 9               303.725          518.3175
 11              272.475          562.6925
 13              241.225          607.0675
 15              209.975          651.4425

Another calculation (Y/N)? N
```

Program notes

(1) Lines 10–30 set up title
(2) Lines 40–70 input information about system
(3) Line 80 provides values of yield coefficient and endogenous respiration coefficient (change these values if required)
(4) Line 90 calculates COD removal
(5) Line 100 calculates non-biodegradable VSS
(6) Lines 110–140 print table heading
(7) Line 150 sets up loop for retention times of 1 to 25 hours at 2-hour intervals (change these values if required)
(8) Line 160 calculates VSS accumulation in time H
(9) Line 170 calculates oxygen requirement in time H
(10) Line 180 prints results
(11) Line 190 returns loop
(12) Lines 200–220 offer another calculation if desired

Example 7.2 KINCOF: biological kinetic coefficients

A method for the experimental determination of kinetic coefficients depends on the use of a series of batch activated-sludge reactors without sludge recycle. The reactors have the same initial BOD and VSS concentration but are operated for different periods of time. Because the reactors do not have sludge recycle, $\theta_c = \theta$. The experimental data must first be put in the form of Equation 7.11 so that a plot of $X\theta/(S_0 - S)$ against $1/S$ can be used to obtain Y/μ_m (the intercept) and YK_s/μ_m (the slope). Hence K_s can be obtained. The data can also be plotted in the form of Equation 7.14 to obtain the value of b (– intercept) and Y (the slope).

A typical set of data would be:

Reactor no.	S_0 (mg/l)	S (mg/l)	$\theta(=\theta_c)$ (day)	X (mg/l)
1	300	7	3.2	130
2	300	13	2.0	130
3	300	18	1.6	130
4	300	31	1.2	130
5	300	39	1.0	130

```
10 CLS
20 REM KINCOF
30 PRINT"Biological Kinetic Coefficients"
40 X=0:Y=0:X2=0:Y2=0:XY=0:W=0:Z=0:W2=0:Z2=0:WZ=0
50 INPUT"Enter initial BOD (mg/l), MLVSS (mg/l)";SI,M
60 INPUT"Enter number of reactors";N
70 FOR R=1 TO N
80 INPUT"Enter retention time (d), final BOD (mg/l)";T,SF
90 A=1/SF
100 E=M*T/(SI-SF)
110 X=X+A
120 Y=Y+E
130 X2=X2+A^2
140 Y2=Y2+E^2
150 XY=XY+A*E
160 C=1/E
170 D=1/T
180 W=W+C
190 Z=Z+D
200 W2=W2+C^2
210 Z2=Z2+D^2
220 WZ=WZ+C*D
230 NEXT R
240 S1=(N*XY-Y*X)/(N*X2-X^2)
250 I1=(Y-S1*X)/N
260 K1=1/I1
270 KS=S1*K1
280 PRINT
290 PRINT"Half vel constant (Ks)=";KS;"mg/l"
300 S2=(N*WZ-Z*W)/(N*W2-W^2)
310 I2=(Z-S2*W)/N
320 B=-I2
330 PRINT"Yield coefficient (Y)=";S2
340 PRINT"Endogenous respiration coeff (b)=";B;"/day"
350 PRINT"Maximum specific growth rate (um)=";K1*S2;"/day"
360 END
```

```
Biological Kinetic Coefficients
Enter initial BOD (mg/l), MLVSS (mg/l)? 300,130
Enter number of reactors? 5
Enter retention time (d), final BOD (mg/l)? 3.2,7
Enter retention time (d), final BOD (mg/l)? 2,13
Enter retention time (d), final BOD (mg/l)? 1.6,18
Enter retention time (d), final BOD (mg/l)? 1.2,31
Enter retention time (d), final BOD (mg/l)? 1,39

Half vel constant (Ks)= 24.84394 mg/l
Yield coefficient (Y)= .5286157
Endogenous respiration coeff (b)= 7.489948E-02 /day
Maximum specific growth rate (um)= 1.694972 /day
```

Program notes

(1) Lines 10–30 set up title
(2) Line 40 initializes variables for least squares
(3) Lines 50–60 enter initial data for reactors
(4) Line 70 sets up loop for reactors
(5) Line 80 enters individual reactor data
(6) Lines 90–150 least squares for Equation 7.11
(7) Lines 180–220 least squares for Equation 7.14
(8) Line 230 returns loop
(9) Lines 240–270 evaluate slope and intercept for Equation 7.11 and calculate K_s
(10) Lines 280–290 print result
(11) Lines 300–320 evaluate slope and intercept for Equation 7.14 and calculate b
(12) Lines 330–350 print results

Example 7.3 ANAEROBE: anaerobic solids and gas production

Write a program to calculate the VSS production and volume of methane released from an anaerobic reactor.

In an anaerobic reactor the VSS in the effluent and the non-biodegradable VSS in the feed are often ignored, so that the accumulation of VSS can be expressed as

$$\text{VSS/unit time} = Y(S_0 - S)q - bXq\theta \qquad (7.20)$$

where q = rate of flow

However, $bXq\theta$ can be replaced by VSS/unit time × θ_c so that

$$\text{VSS/unit time} = Y(S_0 - S)/(1 + b\theta_c) \qquad (7.21)$$

There is, of course, a problem in this context; to calculate the VSS accumulation from Equation 7.20, it is necessary to know the value of θ_c, which is itself a function of the VSS in the system and the daily VSS accumulation. A solution must therefore use an iterative approach, for which a computational method is ideal.

A wastewater with COD 5000 mg/l is to be treated in an anaerobic system to give an effluent of 2000 mg/l COD. The daily flow of wastewater is 10 m^3 and the reactor has a capacity of 25 m^3 with an estimated MLVSS of 10 000 mg/l. Determine the VSS accumulation, solids retention time and methane yield.

```
10 CLS
20 REM ANAEROBE
30 PRINT"Anaerobic Treatment Solids and Gas Production"
40 INPUT"Enter wastewater flow (cu.m/d)";Q
```

```
50 INPUT"Enter waste, and effluent COD values (mg/l)"; C,CE
60 INPUT"Enter reactor volume (cu.m)";RV
70 INPUT"Enter reactor MLVSS (mg/l)";ML
80 Y=.1:B=.01
90 CR= C-CE
100 TS=5
110 VS=Y*CR*Q/(1+B*TS)
120 TM=(RV*ML)/VS
130 OL=(C*Q)*.001/RV
140 DT=RV/Q
150 IF OL > 15 THEN 280
160 IF (TM-TS) < 1 THEN 180
170 TS=TM:GOTO 110
180 PRINT"Volatile solids production=";VS*.001;"kg/d"
190 PRINT"Solids retention time=";TM;"d"
200 PRINT"Hydraulic retention time=";DT;"d"
210 PRINT"Organic loading=";OL;"kg COD/cu.m d"
220 CH=.35*(CR*Q-1.42*VS)*.001
230 PRINT"Methane production=";CH;"cu.m/d"
240 PRINT
250 INPUT"Another calculation (Y/N)";Q$
260 IF Q$="Y" OR Q$="y" THEN 10
270 END
280 CLS
290 PRINT"WARNING - LOADING TOO HIGH - FAILURE LIKELY"
300 PRINT"Organic loading=";OL;"kg COD/cu.m d"
310 PRINT"Should not exceed 15 kg COD/cu.m d"
320 PRINT
330 GOTO 250
340 END
```

```
Anaerobic Treatment Solids and Gas Production
Enter wastewater flow (cu.m/d)? 10
Enter waste, and effluent COD values (mg/l)? 5000,2000
Enter reactor volume (cu.m)? 25
Enter reactor MLVSS (mg/l)? 10000
Volatile solids production= .5043619 kg/d
Solids retention time= 495.6759 d
Hydraulic retention time= 2.5 d
Organic loading= 2 kg COD/cu.m d
Methane production= 10.24933 cu.m/d

Another calculation (Y/N)? N
```

Program notes

(1) Lines 10–30 set up title
(2) Lines 40–70 input information about system and waste
(3) Line 80 provides values of yield coefficient and endogenous respiration coefficient (change these values if required)
(4) Line 90 calculates COD removal
(5) Line 100 fixes initial θ_c at a value which is less than that likely in practice
(6) Line 110 calculates VSS accumulation on basis of given value of θ_c
(7) Line 120 calculates modified θ_c using the value of VSS accumulation from 110

(8) Lines 130–140 calculate organic loading rate and θ
(9) Line 150 warns if organic loading too high and transfers program to 290
(10) Line 160 checks for closeness of calculated θ_c to given value; if difference is < 1 the result is accepted and program passes to results in 180.
(11) Line 170 if calculated θ_c is $>$ initial $\theta_c + 1$ then new value of θ_c is returned to 110 for repeat calculation
(12) Lines 180–210 print results
(13) Lines 220–230 calculate and print methane production
(14) Lines 240–260 offer another calculation if desired
(15) Line 270 stops program running into warning
(16) Lines 280–330 warning of high loading rate

Example 7.4 OXPOND: facultative oxidation pond loadings

A number of empirical relationships have been produced to express the organic loadings which may be applied to facultative oxidation ponds. These are based on ambient temperature or latitude as convenient measures of the local environmental conditions.

The Pescod and McGarry formula for allowable surface loading is

$$\text{kg BOD/ha day} = 60.3 \times 1.0993^{T} \qquad (7.22)$$

where T = minimum ambient mean monthly air temperature

The Mara modification of Equation 7.22 is

$$\text{kg BOD/ha day} = 20T - 120 \qquad (7.23)$$

and the Arthur modification of Equation 7.23 is

$$\text{kg BOD/ha day} = 20T - 60 \qquad (7.24)$$

Based on Indian conditions, Arceivala has developed a relationship which uses the latitude of the site (between 8°N and 36°N) to give

$$\text{kg BOD/ha day} = 375 - 6.25 \times \text{latitude} \qquad (7.25)$$

A program using these relationships can be written to allow the user to input temperature and latitude information and obtain a comparison of the loadings allowed by the various relationships.

```
10 CLS
20 REM OXPOND
30 PRINT"Facultative oxidation pond loadings"
40 INPUT"Enter min mean monthly air temperature (deg C)";T
```

```
50 INPUT"Enter local latitude (degrees)";L
60 LP=60.3*1.0993^T
70 LM=20*T-120
80 LA=20*T-60
90 IF L<8 OR L>36 THEN 110
100 LR=375-6.25*L
110 PRINT
120 PRINT"Design loadings for facultative ponds to give"
130 PRINT"approx 80% BOD removal from domestic sewage"
140 PRINT
150 PRINT"Pescod & McGarry relationship: kg BOD/ha.d=";LP
160 PRINT"Mara relationship: kg BOD/ha.d=";LM
170 PRINT"Arthur relationship: kg BOD/ha.d=";LA
180 IF L<8 OR L>36 THEN 200
190 PRINT"Arceivala relationship: kg BOD/ha.d=";LR
200 PRINT
210 INPUT"Another calculation (Y/N)";Q$
220 IF Q$="Y" OR Q$="y" THEN 10
230 END
```

```
Facultative oxidation pond loadings
Enter min mean monthly air temperature (deg C)? 25
Enter local latitude (degrees)? 30

Design loadings for facultative ponds to give
approx 80% BOD removal from domestic sewage

Pescod & McGarry relationship: kg BOD/ha.d= 643.0185
Mara relationship: kg BOD/ha.d= 380
Arthur relationship: kg BOD/ha.d= 440
Arceivala relationship: kg BOD/ha.d= 187.5

Another calculation (Y/N)? N
```

Program notes

(1) Lines 10–30 set up title
(2) Lines 40–50 enter location details
(3) Lines 60–80 calculate pond loadings for Pescod and McGarry, Mara, and Arthur
(4) Line 90 checks for applicable latitude, else 110
(5) Line 100 calculates loading by Arceivala
(6) Lines 110–140 set up heading
(7) Lines 150–170 print results
(8) Lines 180–190 print Arceivala result if appropriate
(9) Lines 200–220 offer another calculation if desired

PROBLEMS

(7.1) Important control measures for an activated sludge plant are: Sludge Volume Index, Sludge Density Index and Mean Cell Residence Time. Develop a program which will calculate these indices when the basic dimensions and operational analyses of the plant are provided by the user.

(7.2) Develop a program which will enable the operator of an activated-sludge plant to calculate the sludge wastage rate needed to maintain a range of mixed liquor suspended solids concentrations under steady-state loading conditions.

(7.3) Write a program which calculates the organic and hydraulic loading for a biological filter. The user provides dimensions of the filter and flow and BOD data for the wastewater. The program should provide warnings if the loadings calculated exceed maximum values of 0.15 kg BOD/m^3/day and/or 0.75(m^3/m^3)/day.

FURTHER READING

Institute of Water Pollution Control, *Activated Sludge*, IWPC, 1987.

Institution of Water and Environmental Management, *Biological Filtration*, IWEM, 1988

Tebbutt, T. H. Y., *Principles of Water Quality Control*, 3rd edn, Chapters 14 and 15, Pergamon Press, 1983.

Chapter 8

Sludge handling and treatment

In many water and wastewater treatment operations, impurities are removed as settleable solids or as mechanically filtered solids, resulting in the production of sludges. In the case of water treatment, the volumes of sludge are not usually large and their composition is such that the material is relatively inert. With wastewater treatment, however, the situation is very different, with the production of large volumes of often offensive sludges. In sewage treatment, for example, the cost of treating and disposing of the high organic content sludges which result from primary and secondary treatment usually accounts for about half of the overall cost of the operation. A significant feature of almost all water and wastewater treatment sludges is their low solids content, commonly 1–5%. This means that a relatively small volume of solids is associated with very large volumes of liquid, which complicates handling and disposal of the sludge.

ESSENTIAL THEORY

8.1 Properties of sludges

The main types of sludges encountered in water and wastewater treatment are:

1. Organic sludges from primary and secondary settlement of sewage.
2. Anaerobically digested sewage sludges.
3. Inorganic sludges from coagulation of water and certain industrial wastewaters.
4. Inorganic sludges from precipitation softening and precipitation treatment of certain industrial wastewaters.

For sludges from the treatment of domestic sewage, the following quantities may be used as guidelines in the absence of more specific information:

primary sludge 0.034 kg/person day

secondary sludge 0.012 kg/person day

Anaerobic digestion of these sludges will reduce the above values by about 35%.

In the case of coagulation of raw water, an estimate of sludge production can be obtained from:

$$\text{mg/l sludge solids} = \text{SS in raw water} + 0.07 \times {}^{\circ}\text{H colour removed} + \text{hydroxides precipitated} + \text{any other insoluble additives} \quad (8.1)$$

With precipitation reactions the sludge production can be calculated from the chemistry of the process as shown in Section 6.2.

The density and nature of the solids in a sludge have a considerable influence on its characteristics. Thus an inorganic sludge composed of precipitated calcium carbonate pellets will drain readily to give a material containing around 50% solids. Most other sludges are much more difficult to dewater because the solids are usually compressible flocculent materials. These have a great deal of water physically trapped in their structure in addition to water which is chemically bound to the solids. Thus drainage by gravity or induced dewatering by pressure or vacuum require careful control to avoid the development of a compressed layer of solids which becomes effectively impermeable and prevents further removal of liquid. In this respect, homogeneous sludges from biological treatment of wastewater and coagulation of water are more difficult to dewater than a heterogeneous primary settlement sludge in sewage treatment.

The specific gravity of a sludge can be approximated by the following expression:

$$\text{sludge SG} = 100/[(\%\ \text{sludge solids}/\text{solids SG}) + (\%\ \text{water}/\text{water SG})] \quad (8.2)$$

The importance of reducing the moisture content of a sludge can be illustrated by an example. Consider a sludge having 2% solids, the solids having a specific gravity of 1.35. Thus a tonne of sludge would contain 20 kg of solids and 980 kg of water.

$$\text{sludge SG} = 100/[(2/1.35) + (98/1)] = 1.005$$

$$\text{volume occupied by 1 tonne at 2\% solids} = 1000/1.005 \times 1000 = 0.995\ \text{m}^3$$

If dewatered to 30% solids (a soil-like consistency), the 20 kg of solids would then be accompanied by only 66.67 kg of water

$$\text{sludge SG} = 100/[(30/1.35) + (70/1)] = 1.084$$

$$\text{volume occupied by 86.67 kg at 30\% solids} = 86.87/1.084 \times 1000 = 0.080\,\text{m}^3$$

Thus the volume of sludge dewatered to 30% solids is about 8% of the volume occupied at 2% solids. Such reductions in volume are of great importance in sludge handling and disposal operations, particularly in those which require transport of the sludge to a distant site.

Because of the importance of reducing water content, sludges with poor dewatering characteristics cause problems and it is often necessary to condition the sludge to obtain a more effective release of liquid. Gravity thickening of most sludges will give a significant increase in solids content and this can be achieved by slow-speed stirring in a tank with a retention time of around 24 hours. Such treatment will usually give a 50% increase in solids content. With some sludges fine solids hinder the release of water and in such cases chemical coagulation will help to agglomerate these particles and also release some of the water bound to the solids. Common coagulants are lime or aluminium chloride and the addition of small doses of polyelectrolytes can be helpful.

8.2 Measurement of dewatering characteristics

It is clearly of some importance to be able to assess the dewatering characteristics of a sludge for both design and operational purposes. The flow of water through sludge in a dewatering situation bears some similarity to the conditions in a bed of porous media which were discussed in Section 5.3. With this in mind, Coackley modified the Carman–Kozeny filtration equation to provide a means of analysing laboratory dewatering data obtained using the apparatus shown diagrammatically in Figure 8.1. A sample of sludge is placed in the filter funnel and a constant vacuum applied, and the water released from the sludge is collected in a measuring cylinder with the volume accumulated being recorded at known time intervals. The rate of dewatering is a function of the sludge characteristics and the vacuum pressure applied so that

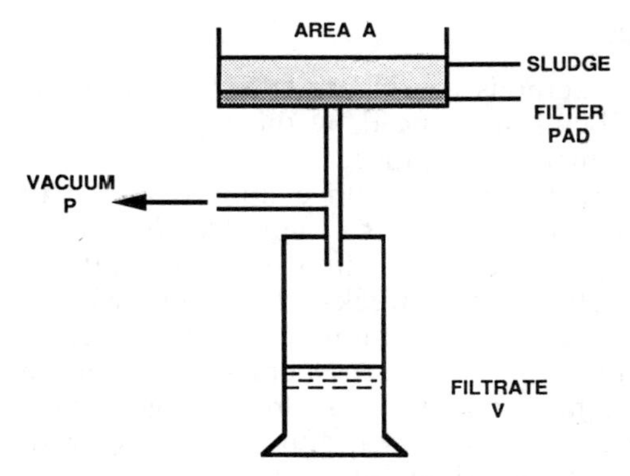

Figure 8.1 Specific resistance to filtration apparatus

$$\mathrm{d}V/\mathrm{d}t = PA^2/\mu(rcV + RA) \tag{8.3}$$

where V = volume of filtrate collected in time t
P = vacuum pressure (standard value is 49 kPa)
A = area of filter unit
μ = absolute viscosity of filtrate
r = specific resistance to filtration of sludge
c = solids content of sludge
R = resistance to filtration of filter unit

At constant pressure, integration of Equation 8.3 gives

$$t = (\mu rc/2PA^2)V^2 + (\mu R/PA)V \tag{8.4}$$

or

$$t/V = (\mu rc/2PA^2)V + \mu R/PA \tag{8.5}$$

Plotting the experimental data in the form of Equation 8.5 enables calculation of the specific resistance of the sludge from the slope of the line. Specific resistance has units of m/kg and a sludge with good dewatering characteristics would have an r value of around 10^{10} m/kg. Poorly dewatering sludges might have r values of 10^{12} or higher.

A simplified method of assessing dewatering characteristics is the Capillary Suction Time (CST), which measures the time taken for water from a small sample of sludge to travel a known distance through a standard absorbent paper. An easily dewatered sludge will have a CST of around 50 seconds and poor sludges may have CSTs of hundreds of seconds.

8.3 Dewatering methods

A variety of dewatering methods are in use, their applicability in a particular situation being influenced by the type of sludge and the ultimate disposal method employed.

Drying beds utilize gravity drainage and evaporation due to the effects of sun and wind and, if depths of liquid sludge of not more than about 250 mm are used, will often produce a liftable cake of 25–30% solids after a few weeks. This time will be much increased in poor weather conditions and this, together with the large area required (about 0.25 m^2/person), has caused drying beds to fall from favour in the UK for sewage sludge dewatering. They can, however, be very useful in countries with warm climates, low land values and labour costs.

In urban areas sewage sludges normally require some form of mechanical dewatering, and for this to be economic, thickening and chemical conditioning are essential. Although both vacuum and pressure filtration can be employed, pressure filtration on a plate or belt filter is usually preferred because of the ease with which cakes of more than 30% solids can be obtained.

8.4 Sludge disposal

The volume of liquid sludge produced in sewage treatment is about 1% of the volume of sewage treated, so that in the UK some 30 million tonnes of sewage sludge must be disposed of each year. Sludge production from water treatment operations is much lower; in the UK the annual production is about one million tonnes. Thus the major problem relates to the disposal of the organic sewage sludges which are highly putrescible and are likely to contain pathogenic microorganisms if not anaerobically digested. The organic content of sewage sludges and the presence of nitrogen and phosphorus compounds do, however, mean that these sludges have some value as soil conditioners. Waterworks sludges, being largely inorganic in nature, have little value but they are relatively inert and are usually tipped on a suitable landfill site. Precipitation sludges from softening can be useful for application to acidic soils.

Because of the soil-conditioning value of sewage sludges the most popular disposal method in the UK is application of undewatered digested sludge to agricultural land. This may not be possible because of the absence of nearby land or because industrial wastes, such as heavy metals, make the sludge unacceptable for land application. Sanitary landfilling and tipping is then a possible option, provided that there is no risk of groundwater contamination. Incineration is usually seen as a last resort because of its cost and possible air pollution problems. In the UK large volumes of liquid sludge are dumped at sea and although there is no clear evidence of environmental damage arising from this disposal method it seems likely that such an option will not be permitted in the future.

WORKED EXAMPLES

Example 8.1 SLUDGE: sludge volume calculations

Write a program which will tabulate sludge specific gravities and volumes for a given weight of sludge over a range of solids contents up to a limit specified by the user.

A typical sewage sludge would have an initial solids content of 4% with a solids specific gravity of 1.4. Determine the specific gravity of the sludge and the volume occupied as 1 tonne of the raw sludge is dewatered to 40% in ten steps.

```
10 CLS
20 REM SLUDGE
30 PRINT"Sludge Volume Calculation"
40 INPUT"Enter initial wt (tonne), solids (%)";M1,S1
50 INPUT"Enter specific gravity of solids";SG
60 INPUT"Enter maximum solids content (%) required";S2
70 WS=S1*M1
80 PRINT
90 PRINT"Solids","Sludge","Volume"
100 PRINT" (%)   "," SG  ","(cu.m)"
110 FOR S=S1 TO S2 STEP (S2-S1)/10
120 SM=100/((S/SG)+100-S)
130 V=WS/(S*SM)
140 PRINT S,SM,V
150 NEXT S
160 PRINT
170 INPUT"Another calculation (Y/N)";Q$
180 IF Q$="Y" OR Q$="y" THEN 10
190 END
```

```
Sludge Volume Calculation
Enter initial wt (tonne), solids (%)? 1,4
Enter specific gravity of solids? 1.4
Enter maximum solids content (%) required? 40

Solids          Sludge          Volume
 (%)             SG             (cu.m)
 4               1.011561        .9885714
 7.6             1.022196        .5148873
 11.2            1.033058        .3457143
 14.8            1.044153        .2588417
 18.4            1.055489        .2059628
```

```
22            1.067073        .1703896
25.6          1.078915        .1448214
29.2          1.091022        .1255577
32.8          1.103405        .1105227
36.4          1.116071        9.846156E-02
40            1.129032        8.857144E-02

Another calculation (Y/N)? N
```

Program notes
(1) Lines 10–30 set up title
(2) Lines 40–60 enter information about sludge
(3) Line 70 calculates weight of solids
(4) Lines 80–100 print table headings
(5) Line 110 sets up loop for solids contents
(6) Lines 120–140 calculate and print SG and volume
(7) Line 150 returns loop
(8) Lines 160–180 offer another calculation if desired

Example 8.2 SPECREST: specific resistance to filtration

Determination of specific resistance involves the measurement of drained liquid at known intervals after application of the standard vacuum. Write a program which will calculate the specific resistance by means of a least squares fit to the experimental data. The program should also calculate the correlation coefficient for the linear fit and warn if correlation is weak.

A typical set of data would be:

Time (s)	*Filtrate volume* (ml)
60	1.4
120	2.4
240	4.2
480	6.9
900	10.4

Vacuum pressure: 49 kPa
Filtration viscosity: 1.011×10^{-3} N s/m^2
Solids concentration: 7.5%
Filter area: 4.42×10^{-3} m^2

```
10 CLS
20 REM SPECREST
30 PRINT"Specific Resistance to Filtration"
40 X=0:Y=0:X2=0:Y2=0:XY=0
50 INPUT"How many data points";N
```

```
60 FOR D=1 TO N
70 INPUT"Enter volume (ml), and time (s)";V,T
80 S=T/V
90 X=X+V
100 Y=Y+S
110 X2=X2+V^2
120 Y2=Y2+S^2
130 XY=XY+V*S
140 NEXT D
150 SL=(N*XY-Y*X)/(N*X2-X^2)
160 PRINT
170 PRINT"Slope of graph=";SL;"sec/ml squared"
180 X=SL*(XY-X*Y/N)
190 Y2=Y2-Y^2/N
200 Y=Y2-X
210 XY=X/Y2
220 CC=SQR(XY)
230 PRINT"Linear correlation coefficient for data=";CC
240 IF CC<.8 THEN PRINT"Low correlation coeff - poor data!"
250 PRINT
260 INPUT"Solids (%), filtrate viscosity (Ns/sq.m)";SS,MU
270 INPUT"Vacuum pressure (kPa), filter area (sq.m)";P,A
280 SR=2*P*1000!*A^2*SL*.000001*1E+18/(MU*SS*10)
290 PRINT"Specific resistance=";SR;"m/kg"
300 PRINT
310 INPUT"Another set of calculations (Y/N)";Q$
320 IF Q$="Y" OR Q$="y" THEN 10
330 END
```

```
Specific Resistance to Filtration
How many data points? 5
Enter volume (ml), and time (s)? 1.4,60
Enter volume (ml), and time (s)? 2.4,120
Enter volume (ml), and time (s)? 4.2,240
Enter volume (ml), and time (s)? 6.9,480
Enter volume (ml), and time (s)? 10.4,900

Slope of graph= 4.728034 sec/ml squared
Linear correlation coefficient for data= .9987016

Solids (%), filtrate viscosity (Ns/sq.m)? 7.5,1.011E-3
Vacuum pressure (kPa), filter area (sq.m)? 49,4.42E-3
Specific resistance= 1.19382E+14 m/kg

Another set of calculations (Y/N)? N
```

Program notes

(1) Lines 10–30 set up title
(2) Line 40 initializes variables for least squares
(3) Line 50 enters number of data points
(4) Line 60 sets up loop for entry of data
(5) Line 70 enters data
(6) Lines 80–130 least squares calculations
(7) Line 140 returns loop
(8) Lines 150–170 calculate and print slope of line
(9) Lines 180–230 calculate and print correlation coefficient
(10) Line 240 warns of weak correlation
(11) Lines 250–270 enters other factors for calculation

(12) Lines 280–290 calculate and print specific resistance
(13) Lines 300–320 offer another calculation if desired

PROBLEMS

(8.1) Modify Worked Example 8.1 to enable the user to obtain the amount of water which must be removed from a sludge to reach a specified solids content.

(8.2) Write a program which will estimate the daily weight of sludge solids produced in a conventional sewage treatment plant. The user will input data for the flow to be treated, its BOD and SS concentrations, and the effluent SS limit. The SS removal in primary sedimentation can be specified and the biological sludge production can be taken as 0.5 kg/kg BOD removed. By specifying sludge solids contents the user will also be given daily volumes of sludge production.

FURTHER READING

Institute of Water Pollution Control, *Sewage Sludge*, Volumes I, II and III, IWPC, 1978, 1979, 1981.
Tebbutt, T. H. Y., *Principles of Water Quality Control*, 3rd edn, Chapter 18, Pergamon Press, 1983.

Chapter 9

Concepts in design

As will be apparent from earlier chapters, most water and wastewater treatment plants consist of a number of unit processes or operations in combination. Each process or operation is intended to carry out a particular function, e.g. sedimentation to remove settleable solids, coagulation and flocculation to convert colloidal solids into settleable solids, biological systems to remove BOD. The selection of an appropriate treatment chain depends upon the performance required and the characteristics of the water or wastewater.

ESSENTIAL THEORY

9.1 Treatment plant design

Since waters and wastewaters vary considerably in composition it is not possible to produce a standard water treatment or wastewater treatment plant which will be suitable for all situations. However, for common situations it is usual to utilize conventional design criteria, at least at the initial stage of the design procedure.

9.1.1 Water treatment

The object of water treatment is to provide a wholesome supply of water which meets the appropriate national or international quality standards. In the European Community the relevant standards are those in the EC Directive 80/778/EEC: *Quality of Water Intended for Human Consumption*; in the USA standards are set out in the Safe Drinking Water Act. The World Health Organization publishes *Guidelines for Drinking Water Quality* which are used as a basis for many national standards. The treatment necessary to achieve the desired water quality will be a function of the quality of the raw water; for a deep groundwater, treatment would probably only be precautionary disinfection, whereas a polluted lowland river source would almost certainly require a complex treatment system. In some cases, particularly when large plants are to be constructed, pilot-scale studies are often well worthwhile and they should certainly be undertaken if conditions are at all unusual.

A common situation is that where water is abstracted from a lowland river flowing through essentially rural surroundings. Such a water will usually be of reasonable quality but it will contain colloidal solids, soluble organic matter and significant numbers of microorganisms. To obtain a satisfactory drinking water from such a source it would normally be necessary to provide coagulation and flocculation, sedimentation, rapid filtration and disinfection. Figure 9.1 shows the flow sheet for such a plant together with the conventional design criteria for each main process.

9.1.2 Sewage treatment

The characteristics of sewage are affected by the nature of the sewerage system and by the presence of industrial wastewaters, so that a standard approach must be used with caution. Nevertheless, it is again possible to produce an initial design based on conventional criteria which are appropriate for sewages which are essentially domestic in nature and for sites where full treatment is required. In a separate sewerage system, foul sewage is collected in one system of pipes and conveyed to a treatment plant. Surface run off is collected in a second system of pipes which discharge to a suitable watercourse. With a combined sewerage system, both foul sewage and surface run off are transported in the same pipes. Because the run off from an urban area in even a moderate rainfall is many times greater than the foul sewage or dry weather flow (DWF) from that area, the sewers rapidly become full. For economic reasons it is impracticable to build sewers large enough to carry all the surface water from an area to a treatment plant and, even if this were possible, it would be quite unrealistic to design the treatment plant to treat this large flow which only occurs in wet weather. It is therefore customary with a combined sewerage system to provide storm water overflows at intervals along the sewers to discharge flows in excess of around 6 DWF to a nearby watercourse. The remaining 6 DWF passes to the treatment plant where it all receives preliminary treatment.

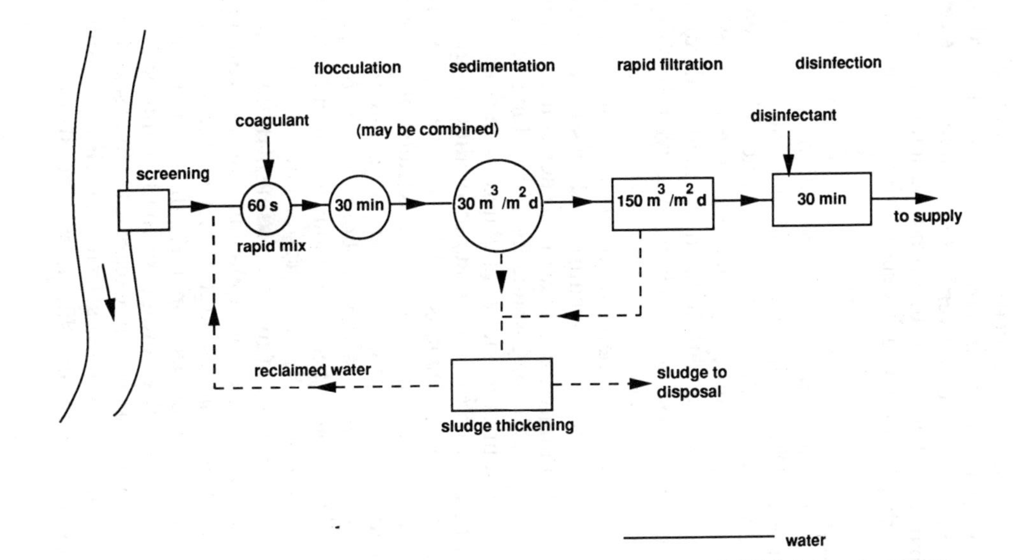

Figure 9.1 Conventional water treatment plant flow sheet

Flows in excess of 3 DWF are then usually diverted to storm water holding tanks and a maximum of 3 DWF passed through the whole plant for full treatment. When the inflow falls below 3 DWF the contents of the storm water tanks are pumped back to the inlet to receive full treatment. In a prolonged or very heavy rainfall the storm water tanks may fill and they are then allowed to discharge directly to a watercourse. On most works such an event will happen on only a few occasions during a year. The discharge from a storm overflow is a mixture of sewage and surface water and is thus likely to pollute the receiving water although, because of the rainfall, this will probably have a higher than normal flow, thus providing additional dilution for the storm overflow discharge. The situation is not ideal but it is based on an economic analysis of the risks associated with pollution which occurs with fairly low probability.

The degree of treatment required at a sewage works is controlled by the effluent consent conditions imposed by the regulatory body. These will usually take into account the use of the receiving water and the dilution available. This river quality objective approach is more rational than the alternative fixed emission concept, which sets a fixed effluent consent without reference to the use of the receiving water or the dilution it provides.

For many situations where there is a reasonable dilution available for the effluent discharge and where the receiving water is not immediately used as a raw water source, an effluent standard of 30 mg/l SS and 20 mg/l BOD has long been used in the UK. More recently, effluent standards have been fixed on a statistical evaluation of the variability in quality of both effluent and receiving water to establish 95% compliance levels. This recognizes the fact that because of the random variations in effluent quality it is impossible to set a standard which will never be exceeded. In most developed countries the per capita contributions of BOD and SS are about 0.055 and 0.080 kg/day respectively. With a DWF of 150 l/person day this gives a domestic sewage strength of around 360 mg/l BOD and 530 mg/l SS. In practice, infiltration into the sewers and cooling water from industrial discharges tend to increase the DWF so that typical sewage strengths in the UK are 250 mg/l BOD and 350 mg/l SS. In the USA sewage strength is lower because of the higher domestic water consumption, and the same may occur in hot climates. A high removal efficiency is thus necessary to achieve a satisfactory effluent quality, so that a combination of processes is essential. Primary sedimentation will usually remove 50–60% of SS and 30–40% of BOD, giving a settled sewage of about 150 mg/l SS and BOD. To obtain an effluent of 30 mg/l SS and 20 mg/l BOD, full biological treatment followed by secondary sedimentation is necessary. Because of the toxic effect of ammonia nitrogen on fish, it is now common to specify a maximum ammonia nitrogen concentration in an effluent of perhaps 5 mg/l. If conditions in the receiving water are critical, it may be appropriate to require an effluent of as low as 5 mg/l of SS, BOD and ammonia nitrogen. This will cause the addition of further processes to the system to provide what is termed tertiary treatment. Figure 9.2 shows the flow sheet and conventional design criteria for a sewage treatment plant intended to produce a 30 : 20 standard effluent from domestic sewage.

9.2 Optimized design

When considering the design of treatment plants it becomes apparent that in many cases the performance capabilities of different processes overlap to at least some extent. For example, in water treatment, reducing the loading on a sedimentation basin will make possible the settlement of smaller particles which would otherwise pass to the filtration stage for removal there. This means that the solids loading on the filters would be reduced so that the filters would operate for a longer period before failing. Since there is a law of diminishing returns for most, if not all, treatment processes, it would be uneconomic to build a very large settling tank so as to remove very small particles. The additional particles removed by sedimentation would probably be capable of removal at the filtration stage at lower cost. This means that the maximum efficiency of the whole treatment system occurs when the performance of the system is optimized. This situation may not, and indeed probably will not, occur when each individual process is operating at its maximum removal capability. The use of systems analysis concepts to develop mathematical optimizing models for treatment plant design has attracted considerable interest over the last few years and is exemplified by the Sewage Treatment Optimizing Model (STOM) developed by CIRIA and WRc in the UK. Similar models have been developed in the USA and elsewhere. Such models can provide useful aids for design, since they enable rapid examination of several alternative treatment plant configurations. It is, however, necessary to appreciate that the value of any model is dependent upon the availability of

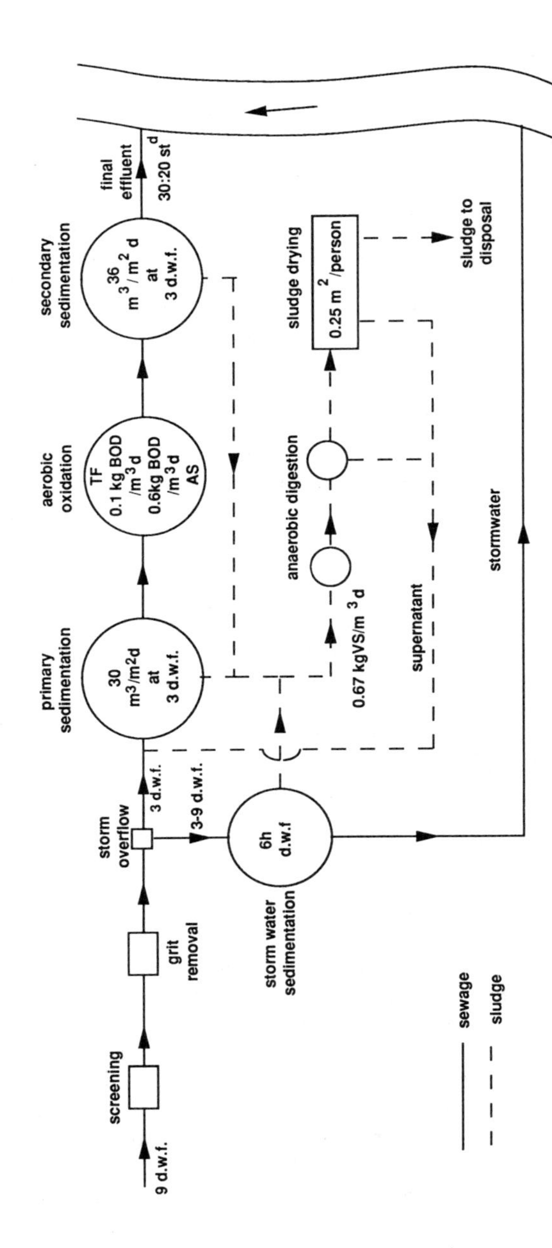

Figure 9.2 Conventional sewage treatment plant flow sheet

reliable performance and cost data for the various process stages. At the present time there is no great difficulty in setting up the actual optimization routines but the collection of realistic input data has not proved to be so easy.

9.2.1 Performance relationships

To provide inputs for an optimizing model it is necessary to establish performance relationships for each treatment process to be incorporated in the model. The concept of a performance relationship is illustrated in Figure 9.3, which shows how the output quality is related to input strength and loading on the process. It is possible to establish such a relationship by using full-scale performance data supplemented by information from pilot-scale and laboratory studies, but the variations in treatability of different waters and wastewaters make precise evaluation of the data difficult. For use in a computer model the performance relationship must be converted into an algorithm form which may be based on theoretical considerations of the treatment process or, more likely, on an empirical relationship.

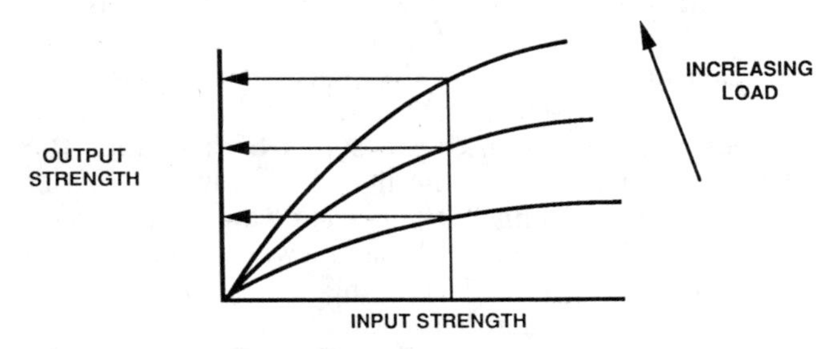

Figure 9.3 Performance relationship

9.2.2 Cost functions

Optimization of a treatment plant implies that the plant achieves the specified performance at the least cost, so it is necessary for the model to be provided with cost data. These must be in the form of cost functions for each individual process stage, as illustrated in Figure 9.4. This information is best obtained by study of the actual construction costs of units in relatively new installations, although care must be taken to allow for any site peculiarities which affect the costs for a particular plant.

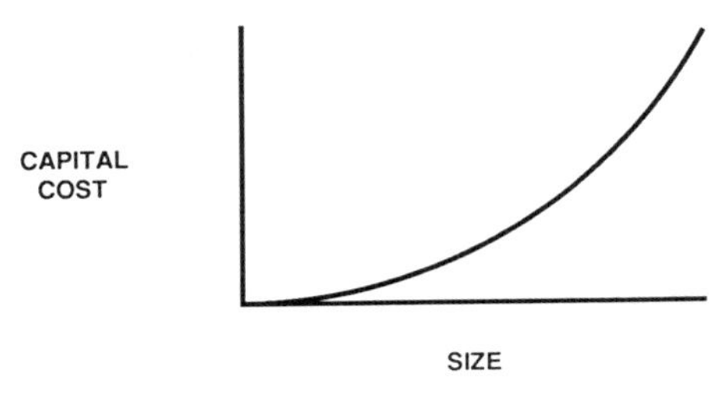

Figure 9.4 Cost function

9.3 Decision-making

An important aspect of a designer's role involves making decisions about many aspects of a project. This could include selection of a water source, choice of appropriate treatment processes, layout of water distribution or sewerage systems, etc. For most of these choices there are likely to be a number of possible options which depend on a variety of other factors. Traditionally, this type of choice can be made easier by setting up some form of decision tree or flow chart which directs the user to the most appropriate solution for a given set of circumstances. Such an aid can take the place of rules committed to memory or shown visually in a diagrammatic format. In practice the choice is made by asking a series of questions the answers of which are considered in the light of knowledge about the system. Thus, when dealing with a wastewater one would ask: are settleable solids present? If the answer is 'yes', sedimentation would be appropriate, but if the answer is 'no', the provision of sedimentation facilities would be inappropriate. This type of decision is, of course, exactly the way in which a computer operates, so that the conversion of a series of questions and answers and the associated rules into a program is ideal for computer-aided decision-making. This is the basis of what are termed 'expert systems' or IKBS (Intelligent Knowledge Based Systems), which are part of the rapidly growing field of IT (Information Technology). Many decisions in real life are not simply 'yes' or 'no' but 'probably yes' or 'possibly yes', so that a true IKBS has to be able to deal with this type of 'fuzzy logic'. In order to build useful IKBSs, languages like BASIC are inappropriate and purpose-made languages such as PROLOG are used. These are able to provide the ability for an IKBS to update its knowledge and learn from mistakes.

WORKED EXAMPLES

Example 9.1 STPDES: sewage treatment plant design

Using the flow sheet and design criteria given in Figure 9.2, write a program which will design the various units in a sewage treatment plant to produce a 30:20 standard effluent from domestic sewage. The user should be able to specify the shape and number of units where appropriate and select the type of biological treatment required. As will be seen from the listing, this program is much longer than the others in this book but it is in fact quite straightforward and once correctly typed it designs a works very quickly.

```
10 CLS
20 REM STPDES
30 PRINT"Sewage Treatment Plant Design"
40 PRINT
50 PRINT"This program uses conventional design parameters"
60 PRINT"to give a preliminary design for a treatment plant"
70 PRINT"producing a 30:20 std effluent from domestic sewage"
80 PRINT
90 INPUT"Enter population, and per capita flow (l/d)";P,PF
100 B=55*1000/PF:S=80*1000/PF
110 INPUT"Enter max storm, full treatment flows (x DWF)";M,T
120 QD=P*PF*.001:QM=QD*M:QT=QD*T
130 PRINT
140 PRINT"Max flow","Max treatd","  DWF  "
150 PRINT" cu.m/d "," cu.m/d   "," cu.m/d"
160 PRINT QM,QT,QD
170 AS=QM/86400!
180 PRINT
190 PRINT"Inlet screens:"
200 PRINT"FACE AREA OF SCREENS=";AS;"sq.m"
210 PRINT"Grit channels:"
220 INPUT"Enter maximum depth of flow for grit channels";DG
230 GA=QM/(.3*86400!)
240 PRINT"Total grit channel X-sectional area=";GA;"sq.m"
250 INPUT"Enter number of grit channels required";GN
260 GW=3*QM/(GN*2*DG*.3*86400!)
270 GL=DG*60*.3*1.5/1.2
280 PRINT GN;"PARABOLIC CHNLS";DG;"m D";GW;"m W";GL;"m L"
290 PRINT
300 INPUT"Try different dimensions (Y/N)";Q$
310 IF Q$="Y"OR Q$="y" THEN 210
320 PRINT
330 PA=QT/30
340 PRINT"Primary sedimentation:"
350 PRINT"Total surface area needed=";PA;"sq.m"
360 INPUT"How many tanks required";A
370 INPUT"Circular or rectangular (C/R)";A$
380 IF A$="R" OR A$="r" THEN 420
390 D=SQR(4*PA/(A*3.14))
400 PRINT A;"PRIMARY SEDIMENTATION TANKS:"D;"m DIAMETER"
410 PRINT" WEIR LOADING=";QT/(A*D*3.14);"cu.m/m.d":GOTO 460
420 INPUT"Enter length:width ratio required";LW
430 W=SQR(PA/(A*LW))
440 PRINT A;"PRIMARY SEDN TKS";LW*W;"m L BY";W;"m W"
450 PRINT" WEIR LOADING=";QT/(A*W);"cu.m/m.d"
```

```
460 INPUT"Try different dimensions (Y/N)";Q$
470 IF Q$="Y" OR Q$="y" THEN 360
480 BE=.7*B:SE=.4*S
490 IF M=T THEN 610
500 PRINT
510 PRINT"Storm tanks:"
520 SV=QD/4
530 SA=SV/2.5
540 PRINT"Storm tank area needed=";SA;"sq.m"
550 INPUT"How many tanks required";ST
560 INPUT"Enter length:width ratio required";LW
570 WS=SQR(SA/(ST*LW))
580 PRINT ST;"STORM TANKS";LW*WS;"m LONG";WS;"m WIDE"
590 INPUT"Try different dimensions (Y/N)";Q$
600 IF Q$="Y" OR Q$="y" THEN 550
610 BL=BE*QD*.001
620 PRINT"Biological treatment:"
630 INPUT"Select bio filter or activated sludge (F/A)";S$
640 IF S$="A" OR S$="a" THEN 770
650 TV=BL/.1:HV=QD/.5
660 PRINT"Filter volume on BOD load basis=";TV;"cu.m"
670 PRINT"Filter volume on hydraulic load basis=";HV;"cu.m"
680 INPUT"Enter selected filter volume";FV
690 FA=FV/1.8
700 PRINT"Filter area needed=";FA;"sq.m"
710 INPUT"How many filters required";NF
720 DF=SQR(4*FA/(NF*3.14))
730 PRINT NF;"BIOLOGICAL FILTERS";DF;"m DIAMETER"
740 INPUT"Try an alternative system (Y/N)";Q$
750 IF Q$="Y" OR Q$="y" THEN 630
760 GOTO 900
770 INPUT"Enter chosen depth of aeration tank (m)";DA
780 AV=BL/.6:AA=AV/DA
790 PRINT"Area of aeration tanks on BOD loading=";AA;"sq.m"
800 VH=QD*1.5*.25:AH=VH/DA
810 PRINT"Area of aeration tanks on volume basis=";AH;"sq.m"
820 INPUT"Enter selected aeration tank area";AD
830 INPUT"How many aeration tanks required";NA
840 A1=AD/NA
850 INPUT"Enter length:width ratio required";AR
860 WA=SQR(A1/AR)
870 PRINT NA;"AERATION TANKS";AR*WA;"m L.";WA;"m W.";DA;"m D"
880 INPUT"Try an alternative system (Y/N)";Q$
890 IF Q$="Y" OR Q$="y" THEN 630
900 PRINT
910 PRINT"Final sedimentation tanks:"
920 FA=QT/36
930 PRINT"Final sedimentation tank area=";FA;"sq.m"
940 INPUT"How many final sedimentation tanks required";NF
950 DF=SQR(4*FA/(NF*3.14))
960 PRINT NF;"FINAL SEDIMENTATION TANKS";DF;"m DIAMETER"
970 INPUT"Try another number of final sedn tanks (Y/N)";Q$
980 IF Q$="Y" OR Q$="y" THEN 940
990 PRINT
1000 PRINT"Sludge treatment units:"
1010 SY=(.7*.6*S+.5*(BE-20))*.001*QD
1020 DV=SY/.67
1030 PRINT"Digester volume needed=";DV;"cu.m"
1040 INPUT"How many digesters required";ND
1050 INPUT"Enter depth:diameter ratio required";DD
1060 DG=(4*DV/(ND*DD*3.142))^.33
1070 PRINT ND;"DIGESTERS";DG;"M DIAMETER";DD*DG;"M DEEP"
1080 INPUT"Try different dimensions (Y/N)";Q$
1090 IF Q$="Y" OR Q$="y" THEN 1040
1100 INPUT"Select filter press or drying bed (P/B)";Q$
1110 IF Q$="B" OR Q$="b" THEN 1160
```

```
1120 INPUT"Enter number of pressings per week":PN
1130 AP=1.5*QD*.001*(.6*S+.5*(BE-20))*.001/PN
1140 PRINT"FILTER PRESS AREA=";AP*1000;"sq.m"
1150 GOTO 1180
1160 AB=P*.25
1170 PRINT"DRYING BED AREA=";AB;"sq.m"
1180 PRINT
1190 INPUT"Display design criteria used (Y/N)";Q$
1200 IF Q$="N" OR Q$="n" THEN 1320
1210 PRINT"DESIGN CRITERIA USED IN PROGRAM"
1220 PRINT
1230 PRINT"Screen face vel=1 m/s:Grit settling vel=0.3 m/s"
1240 PRINT"P sed tank O/F rate=30 m/d, 2 h retn (at 3 DWF)"
1250 PRINT"Storm tank retn=6 h at DWF"
1260 PRINT"Biofilter 0.1 kgBOD/cu.m.d: hyd ldg  0.5 per day"
1270 PRINT"Act sludge 0.6 kgBOD/cu.m.d"
1280 PRINT"Final sed tank O/F rate=36 m/d at 3 DWF"
1290 PRINT"Digester organic ldg=0.67 kgVS/cu.m.d"
1300 PRINT"Drying bed area=0.25 sq.m/person"
1310 PRINT"Press area  WRc TR 61 equation"
1320 PRINT
1330 INPUT"Another design (Y/N)";Q$
1340 IF Q$="Y" OR Q$="y" THEN 10
1350 END
```

```
This program uses conventional design parameters
to give a preliminary design for a treatment plant
producing a 30:20 std effluent from domestic sewage

Enter population, and per capita flow (l/d)? 50000,125
Enter max storm, full treatment flows (x DWF)? 9,3

Max flow        Max treatd        DWF
 cu.m/d          cu.m/d          cu.m/d
 56250           18750           6250

Inlet screens:
FACE AREA OF SCREENS= .6510417 sq.m
Grit channels:
Enter maximum depth of flow for grit channels? 0.75
Total grit channel X-sectional area= 2.170139 sq.m
Enter number of grit channels required? 2
 2 PARABOLIC CHNLS .75 m D 2.170139 m W 16.875 m L

Try different dimensions (Y/N)? N

Primary sedimentation:
Total surface area needed= 625 sq.m
How many tanks required? 4
Circular or rectangular (C/R)? C
 4 PRIMARY SEDIMENTATION TANKS: 14.10832 m DIAMETER
 WEIR LOADING= 105.8124 cu.m/m.d
Try different dimensions (Y/N)? N

Storm tanks:
Storm tank area needed= 625 sq.m
How many tanks required? 3
Enter length:width ratio required? 6
 3 STORM TANKS 35.35534 m LONG 5.892557 m WIDE
Try different dimensions (Y/N)? N
Biological treatment:
Select bio filter or activated sludge (F/A)? F
```

```
Filter volume on BOD load basis= 19250 cu.m
Filter volume on hydraulic load basis= 12500 cu.m
Enter selected filter volume? 12500
Filter area needed= 6944.445 sq.m
How many filters required? 4
 4 BIOLOGICAL FILTERS 47.02772 m DIAMETER
Try an alternative system (Y/N)? Y

Try an alternative system (Y/N)? Y
Select bio filter or activated sludge (F/A)? A
Enter chosen depth of aeration tank (m)? 5
Area of aeration tanks on BOD loading= 641.6667 sq.m
Area of aeration tanks on volume basis= 468.75 sq.m
Enter selected aeration tank area? 500
How many aeration tanks required? 4
Enter length:width ratio required? 8
 4 AERATION TANKS 31.62278 m L. 3.952847 m W. 5 m D
Try an alternative system (Y/N)? N

Final sedimentation tanks:
Final sedimentation tank area= 520.8333 sq.m
How many final sedimentation tanks required? 4
 4 FINAL SEDIMENTATION TANKS 12.87907 m DIAMETER
Try another number of final sedn tanks (Y/N)? N

Sludge treatment units:
Digester volume needed= 3850.747 cu.m
How many digesters required? 3
Enter depth:diameter ratio required? .9
 3 DIGESTERS 11.89821 M DIAMETER 10.70839 M DEEP
Try different dimensions (Y/N)? N
Select filter press or drying bed (P/B)? B
DRYING BED AREA= 12500 sq.m

Display design criteria used (Y/N)? Y
DESIGN CRITERIA USED IN PROGRAM

Screen face vel=1 m/s:Grit settling vel=0.3 m/s
P sed tank O/F rate=30 m/d, 2 h retn (at 3 DWF)
Storm tank retn=6 h at DWF
Biofilter 0.1 kgBOD/cu.m.d: hyd ldg  0.5 per day
Act sludge 0.6 kgBOD/cu.m.d
Final sed tank O/F rate=36 m/d at 3 DWF
Digester organic ldg=0.67 kgVS/cu.m.d
Drying bed area=0.25 sq.m/person
Press area  WRc TR 61 equation

Another design (Y/N)? N
```

Program notes

(1) Lines 10–80 set up titles
(2) Line 90 enters population and flow data
(3) Line 100 calculates per capita BOD and SS values
(4) Line 110 enters flow ratios
(5) Line 120 calculates design flows
(6) Lines 130–160 print design flows
(7) Line 170 calculates screen area at 1 m/s velocity
(8) Lines 180–200 print screen area
(9) Lines 210–280 design grit channels at 0.3 m/s velocity
(10) Lines 290–310 offer another calculation if desired

(11) Lines 320–370 primary settling data for ofr of 30 m/day
(12) Line 380 directs program to skip circular tank
(13) Lines 390–410 circular tank design
(14) Lines 420–450 rectangular tank design
(15) Lines 460–470 offer another calculation if desired
(16) Line 480 calculates load to biological stage
(17) Line 490 skips storm tanks if no storm flow
(18) Lines 500–600 storm tank design for 6 hours capacity
(19) Line 610 calculates BOD load
(20) Lines 620–630 set up titles for biological stage
(21) Line 640 directs program to skip filter
(22) Lines 650–750 filter design based on 0.1 kg BOD/m^3 day or 0.5 m^3/m^3 day and 1.8 m depth
(23) Line 760 directs program to skip activated sludge
(24) Lines 770–890 activated sludge design based on 0.6 kg BOD/m^3 day
(25) Lines 900–980 final settling tank design for ofr 36 m/day
(26) Lines 990–1090 sludge digester design based on loading of 0.67 kg VS/m^3 day VS yield in 1010
(27) Lines 1100–1170 sludge dewatering design
(28) Lines 1180–1310 print design criteria if requested
(29) Lines 1320–1340 offer another design if desired

Example 9.2 OPTDES: demonstration of optimized design

A simple demonstration of an optimized design can be obtained by considering two stages of a sewage treatment plant comprising primary settlement and activated sludge.

The suspended solids removal efficiency of a primary settling tank is assumed to be represented by

$$E = a \exp\left[-\left(\frac{c}{s_i} + bq\right)\right] \tag{9.1}$$

where a, b, and c are constants for a particular wastewater
s_i = influent SS concentration
q = surface overflow rate

The COD removed by sedimentation of organic solids will be a function of E, so that the COD removal efficiency of a primary settling tank can be represented by

$$EL = (d + fE)/[(L_i/s_i) + h] \tag{9.2}$$

where d, f and h are constants for a particular wastewater
L_i = soluble COD concentration in influent

Then the COD of the settled wastewater (L_s) is given by

$$L_s = L_i - ELL_i \tag{9.3}$$

The removal of COD from a particular wastewater is likely to be a function of the MLVSS in the aerator and the hydraulic retention time, so that an expression of the form below can be assumed:

$$L_s/L_e = (1 + nkSt)^{-(1/n)} \tag{9.4}$$

This can be rearranged as

$$t = [(L_s/L_e)^{1/n} - 1]/nkS \tag{9.5}$$

where n and k are constants for a particular wastewater and system
L_e = allowable effluent COD concentration
S = MLVSS concentration
t = hydraulic retention time

Cost functions for the primary and secondary sedimentation tanks can be assumed to be a direct function of the flow treated, and for the aeration tank it is usually assumed that the cost is a power function of the volume required.

It is thus possible to write a program which will determine the sizes and costs of the system over a range of primary tank surface overflow rates. Inspection of the results will indicate the optimum design loading, although for a more complex system a specialized optimizing routine would be employed.

```
10 CLS
20 REM OPTDES
30 PRINT"Demonstration of Optimized Design"
40 PRINT
50 INPUT"Enter influent BOD (mg/l), SS (mg/l)";LI,SI
60 INPUT"Enter design flow (cu.m/s), effl BOD (mg/l)";QD,LE
70 INPUT"Enter lower, upper PST o/f rates (m/d)";QL,QH
80 QQ=(QH-QL)/10
90 INPUT"Enter MLSS in aeration unit (mg/l)";S
100 PRINT
110 PRINT"  Primary Settling Tank          Act Sl   Tot Capl"
120 PRINT"  O/F        SS          BOD       Aern        Cost "
130 PRINT" Rate      Remvl        Remvl      Time        £000 "
140 PRINT" (m/d)      Effy         Effy       (h)             "
150 FOR Q=QL TO QH STEP QQ
160 E=.955*1/(EXP(265/SI+.002*Q))
170 EL=(.31+.78*E)/(200/SI+1.09)
180 LS=LI-EL*LI
190 T=((LS/LE)^1.55-1)/(.000792*S)
200 PC=QD*86.4*225/Q
210 AC=(QD*3.6*T)^.73*200
220 SC=QD*3.6*225
230 TC=PC+AC+SC
240 PRINT TAB(2)Q;TAB(8)E;TAB(19)EL;TAB(30)T;TAB(42)TC
```

```
250 NEXT Q
260 PRINT
270 INPUT"Another set of parameters (Y/N)";Q$
280 IF Q$="Y" OR Q$="y" THEN 10
290 END
```

```
Demonstration of Optimized Design

Enter influent BOD (mg/l), SS (mg/l)? 200,325
Enter design flow (cu.m/s), effl BOD (mg/l)? .1,20
Enter lower, upper PST o/f rates (m/d)? 40,140
Enter MLSS in aeration unit (mg/l)? 2500

  Primary Settling Tank          Act Sl    Tot Capl
  O/F       SS         BOD        Aern       Cost
 Rate     Remvl      Remvl        Time       £000
 (m/d)     Effy       Effy         (h)
  40     .3900702   .3601855   8.463424    580.6653
  50     .3823463   .3566527   8.540298    573.9326
  60     .3747753   .35319     8.615872    570.3821
  70     .3673543   .3497958   8.690164    568.6268
  80     .3600802   .3464688   8.763196    567.9733
  90     .3529501   .3432077   8.834977    568.0369
  100    .3459613   .3400111   8.905531    568.5873
  110    .3391107   .3368779   8.97487     569.478
  120    .3323959   .3338067   9.043012    570.6122
  130    .3258141   .3307963   9.109968    571.9223
  140    .3193625   .3278455   9.175763    573.3611

Another set of parameters (Y/N)? N
```

Program notes

(1) Lines 10–40 set up titles
(2) Lines 50–70 enter basic information for design
(3) Line 80 sets interval for overflow rate values
(4) Lines 90–140 print headings for table
(5) Line 150 sets up loop for each overflow rate
(6) Line 160 Equation 9.1 with appropriate constants
(7) Line 170 Equation 9.2 with appropriate constants
(8) Line 180 Equation 9.3
(9) Line 190 Equation 9.5 with appropriate constants
(10) Lines 200–220 cost functions
(11) Line 230 calculates total cost
(12) Line 240 prints results
(13) Line 250 returns loop
(14) Lines 260–280 offer another calculation if desired

Example 9.3 RURWAT: choice of a water source

A decision chart for selection of a rural water source in a developing country is shown in Figure 9.5. Write a program which takes a user through the decisions to give the recommended source. Such a program is a quasi-expert system which

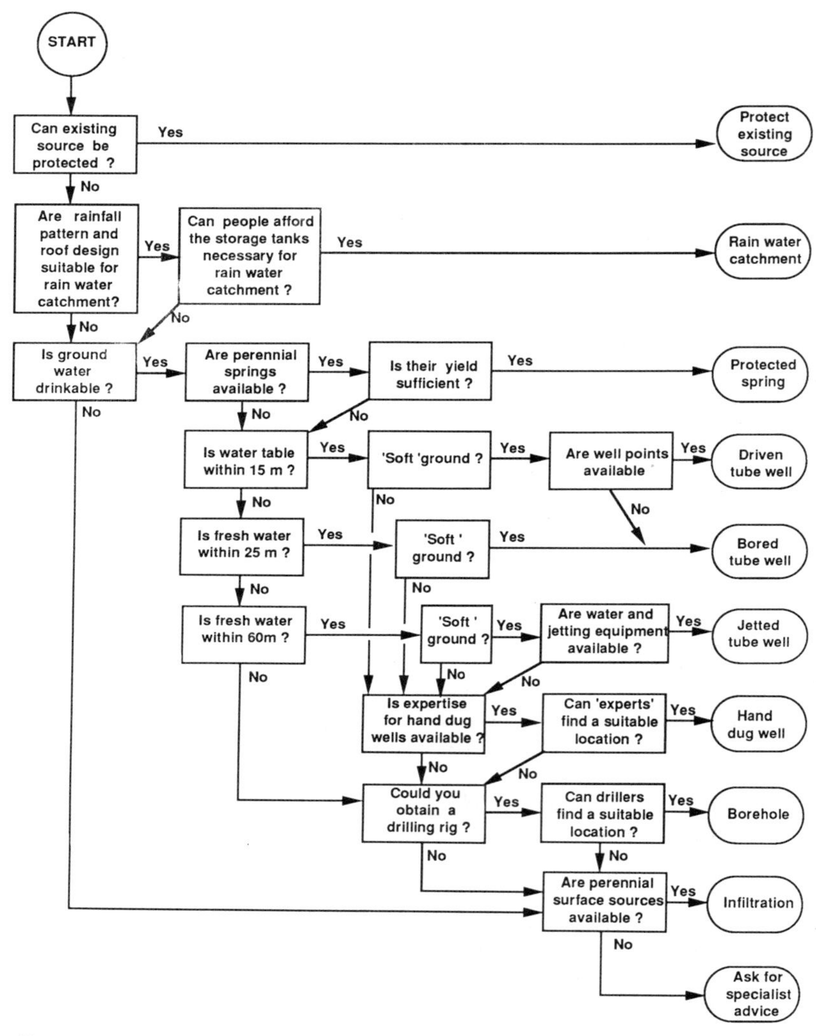

Figure 9.5 Water sources decision chart (from Cairncross, S. and Feachem, R.G., *Small Water Supplies*, Ross Institute, 1978)

uses the 'yes' or 'no' response to questions to direct the flow of the program. The program is basically very simple in content but requires care in ensuring that the user is always directed towards the correct recommendation. In essence the program is just a computerized flow chart, and visual inspection of the flow chart would produce a quicker decision than using the program. The principles used in the program could, however, be utilized for a more complicated decision chart in which a computer solution could be quicker than visual consideration of the alternatives.

```
10 CLS
20 REM RURWAT
30 PRINT"Choice of a Water Source"
40 INPUT"Can existing source be protected (Y/N)";Q$
50 IF Q$="Y"OR Q$="y" THEN 300
60 INPUT"Rnfl ptrn/roof design OK for collectn (Y/N)";Q$
70 IF Q$="Y" OR Q$="y" THEN INPUT"Storage affdble (Y/N)";Q$
80 IF Q$="Y" OR Q$="y" THEN 310
90 INPUT"Is groundwater drinkable (Y/N)";Q$
100 IF Q$="N" OR Q$="n" THEN 320
110 INPUT"Are perennial springs available (Y/N)";Q$
120 IF Q$="Y" OR Q$="y" THEN 350
130 INPUT"Is water table within 15 m (Y/N)";Q$
140 IF Q$="N" OR Q$="n" THEN 180
150 IF Q$="Y" OR Q$="y" THEN INPUT"Is ground soft (Y/N)";Q$
160 IF Q$="Y" OR Q$="y" THEN 370
170 IF Q$="N" OR Q$="n" THEN 460
180 INPUT"Is water table within 25 m (Y/N)";Q$
190 IF Q$="N" OR Q$="n" THEN 230
200 IF Q$="Y" OR Q$="y" THEN INPUT"Is ground soft (Y/N)";Q$
210 IF Q$="Y" OR Q$="y" THEN 420
220 IF Q$="N" OR Q$="n" THEN 460
230 INPUT"Is water table within 60 m (Y/N)";Q$
240 IF Q$="Y" OR Q$="y" THEN 430
250 IF Q$="N" OR Q$="n" THEN 510
300 PRINT"PROTECT EXISTING SOURCE EFFECTIVELY":END
310 PRINT"USE RAIN WATER CATCHMENT":END
320 INPUT"Are perennial surface sources available (Y/N)";Q$
330 IF Q$="Y" OR Q$="y" THEN PRINT"USE INFILTRATION AREA":END
340 IF Q$="N" OR Q$="n" THEN PRINT"GET EXPERT ADVICE":END
350 INPUT"Is their yield sufficient (Y/N)";Q$
360 IF Q$="Y" OR Q$="y" THEN PRINT"USE PROTECTED SPRING":END
365 IF Q$="N" OR Q$="n" THEN 130
370 INPUT"Are well points available (Y/N)";Q$
380 IF Q$="Y" OR Q$="y" THEN PRINT"USE DRIVEN TUBE WELL":END
390 IF Q$="N" OR Q$="n" THEN 420
400 INPUT"Is ground soft (Y/N)";Q$
410 IF Q$="N"OR Q$="n"THEN 460
420 PRINT"USE BORED TUBE WELL":END
430 INPUT"Is ground soft (Y/N)";Q$
440 IF Q$="Y" OR Q$="y" THEN INPUT"Jetting poss (Y/N)";Q$
450 IF Q$="Y" OR Q$="y" THEN PRINT"USE JETTED TUBE WELL":END
460 INPUT"Hand dug well expertize available (Y/N)";Q$
470 IF Q$="N" OR Q$="n" THEN 510
480 INPUT"Is suitable well location available (Y/N)";Q$
490 IF Q$="N" OR Q$="n" THEN 510
500 PRINT"USE HAND DUG WELL":END
510 INPUT"Is drilling rig available (Y/N)";Q$
520 IF Q$="N"OR Q$="n"THEN 320
530 IF Q$="Y" OR Q$="y" THEN INPUT"Site for bore (Y/N)";Q$
540 IF Q$="Y" OR Q$="y" THEN PRINT"USE DRILLED BOREHOLE":END
550 GOTO 320
560 END
```

```
Choice of a Water Source
Can existing source be protected (Y/N)? N
Rnfl ptrn/roof design OK for collectn (Y/N)? N
Is groundwater drinkable (Y/N)? Y
Are perennial springs available (Y/N)? N
Is water table within 15 m (Y/N)? Y
Is ground soft (Y/N)? Y
Are well points available (Y/N)? Y
USE DRIVEN TUBE WELL

Choice of a Water Source
Can existing source be protected (Y/N)? N
Rnfl ptrn/roof design OK for collectn (Y/N)? N
Is groundwater drinkable (Y/N)? Y
Are perennial springs available (Y/N)? N
Is water table within 15 m (Y/N)? N
Is water table within 25 m (Y/N)? N
Is water table within 60 m (Y/N)? Y
Is ground soft (Y/N)? N
Hand dug well expertize available (Y/N)? N
Is drilling rig available (Y/N)? Y
Site for bore (Y/N)? Y
USE DRILLED BOREHOLE
```

Program notes

(1) Lines 10–30 set up title
(2) Line 40 first question
(3) Line 50 directs program to 300 if existing source can be protected and ends run
(4) Line 80 directs program to 310 if rainfall tanks are feasible and ends run
(5) Line 100 directs program to 320 if groundwater is undrinkable
(6) Lines 110–250 aquifer characteristics, response to queries directs program to appropriate queries and decisions in 350–540
(7) Lines 350–540 groundwater queries and decisions
(8) Line 550 reconsiders surface source if groundwater proves to be unsatisfactory

PROBLEMS

(9.1) Write a program to accept experimental data from pilot-scale sand filters, in the form of filter depths and length of run to meet head loss and filtrate turbidity criteria, so that the optimum filter design can be identified.

(9.2) Using the information given in Figure 9.1 and the concepts illustrated in Working Example 9.2, develop a program to undertake the preliminary design of a water treatment plant for a river source.

(9.3) Use the information in Section 6.2 and given in Worked Example 6.2 to develop a decision chart for selection of the most appropriate softening process for waters of different compositions. Use this chart as a basis for a program to present the user with the best process for a particular raw water.

FURTHER READING

Cairncross, S. and Feachem, R. G., *Small Water Supplies*, Ross Institute, 1978.

Feachem, R. G. and Cairncross, S., *Small Excreta Disposal Systems*, Ross Institute, 1978.

Hammer, M. J., *Water and Wastewater Technology*, 2nd edn, Wiley, 1985.

Metcalf and Eddy, *Wastewater Engineering: Treatment Disposal Reuse*, 2nd edn, Tata McGraw Hill, 1979.

Montgomery, J. M., *Water Treatment: Principles and Design*, Wiley, 1985.

Twort, A. C., Law, F. M. and Crowther, F. W., *Water Supply*, 3rd edn, Arnold, 1985.

Water Research Centre, *Cost Information for Water Supply and Sewage Disposal*, WRC, 1977.

Index